LONDON MATHEMATICAL SOCIETY LECTURE NOTE S͏ ͏ ͏

Managing Editor: Professor Endre Süli, Mathematical Institute,
University of Oxford, Woodstock Road, Oxford OX2 6GG, United Kin

The titles below are available from booksellers, or from Cambridge Ur
www.cambridge.org/mathematics

London Mathematical Society Lecture Note Series: 466

Invariance of Modules under Automorphisms of their Envelopes and Covers

ASHISH K. SRIVASTAVA
Saint Louis University

ASKAR TUGANBAEV
National Research University
(Moscow Power Engineering Institute)

PEDRO A. GUIL ASENSIO
Universidad de Murcia

CAMBRIDGE
UNIVERSITY PRESS

CAMBRIDGE
UNIVERSITY PRESS

University Printing House, Cambridge CB2 8BS, United Kingdom

One Liberty Plaza, 20th Floor, New York, NY 10006, USA

477 Williamstown Road, Port Melbourne, VIC 3207, Australia

314-321, 3rd Floor, Plot 3, Splendor Forum, Jasola District Centre, New Delhi – 110025, India

79 Anson Road, #06–04/06, Singapore 079906

Cambridge University Press is part of the University of Cambridge.

It furthers the University's mission by disseminating knowledge in the pursuit of education, learning, and research at the highest international levels of excellence.

www.cambridge.org
Information on this title: www.cambridge.org/9781108949538
DOI: 10.1017/9781108954563

© Cambridge University Press 2021

First published 2021

Printed in the United Kingdom by TJ Books Limited, Padstow Cornwall

A catalogue record for this publication is available from the British Library.

ISBN 978-1-108-94953-8 Paperback

Contents

v

Preface

The study of modules that are invariant under the action of certain subsets of the endomorphism ring of their injective envelope can be drawn back to the pioneering work of Johnson and Wong in which they characterized quasi-injective modules as those modules that are invariant under any endomorphism of their injective envelope. Later, Dickson and Fuller studied modules that are invariant under the group of all automorphisms of their injective envelope and proved that any indecomposable automorphism-invariant module over an $\mathbb{F}$-algebra A is quasi-injective, provided that $\mathbb{F}$ is a field with more than two elements. But after that, this topic remained in dormant stage for some time until Lee and Zhou picked it up again in their paper where they called such modules auto-invariant modules. But the major breakthrough on this topic came from two papers that appeared a few months later. One of them was a paper of Er, Singh and Srivastava, where they proved that the automorphism-invariant modules are precisely the pseudo-injective modules studied earlier by Teply, Jain, Clark, Huynh and others. The other one was a paper by Guil Asensio and Srivastava, where they proved that automorphism-invariant modules satisfy the exchange property and also that they provide a new class of clean modules. Soon after this, Guil Asensio and Srivastava extended the result of Dickson and Fuller by proving that if A is an algebra over a field $\mathbb{F}$ with more than two elements, then a module over A is automorphism invariant if and only if it is quasi-injective.

In 2015, in a paper published in the *Israel Journal of Mathematics*, Guil Asensio, Tutuncu and Srivastava laid down the foundation of general theory of modules invariant under automorphisms (respectively endomorphisms) of envelopes and covers. In this general theory of modules invariant under auto-morphisms (respectively endomorphisms) of envelopes and covers, we have obtained many interesting properties of such modules and found examples of some important classes of modules. When this theory is applied to some particular situations, then we obtain results that extend and simplify several results existing in the literature. For example, as a consequence of these general results, one obtains that modules invariant under automorphisms of their injective (respectively pure-injective) envelopes satisfy the full exchange property. These results extend well-known results of Warfield, Fuchs, Huisgen-Zimmermann and Zimmermann. Most importantly, this study yields us a new tool and new perspective to look at generalizations of injective, pure-injective or flat-cotorsion modules. Until now, most of the generalizations of injective modules were focussed on relaxing conditions on lifting of homomorphisms, but this theory has opened up a whole new direction in the study of module theory.

Chapter 1 presents basic definitions and results in ring theory and module theory that is needed to understand the content of this monograph. In Chapter 2, we give the basics of the general theory of envelopes and covers and introduce the theory of modules invariant under automorphisms of their envelopes. In Chapter 1, we also show the connection of this theory to the additive unit structure of elements in von Neumann regular rings. Chapter 3 presents the decomposition theorem of modules invariant under automorphisms of their envelopes and the structure of endomorphism rings of such modules. In Chapter 4, we discuss the particular case of modules invariant under automorphisms of their injective envelopes. In Chapter 5, we present the dual theory of modules coinvariant under automorphisms of their covers.

In Chapter 6, we study the Schröder–Bernstein problem for modules invariant under automorphism of their envelopes and covers. The Schröder–Bernstein theorem is a classical result in basic set theory. The type of problem where one asks if two mathematical objects A and B that are similar in some sense to part of each other are also similar is usually called the Schröder–Bernstein problem. The question whether Schröder–Bernstein property holds for automorphism-invariant modules was raised by Facchini in one of his papers that was later answered in the affirmative by Guil Asensio, Kaleboğaz and Srivastava. Recently, Tuganbaev developed the theory of automorphism-extendable modules; and, along with his coauthors Abyzov, Quynh and others, he has also developed the theory of automorphism-liftable modules. Chapter 7 presents the theory of automorphism-extendable modules, and Chapter 8 presents the theory of automorphism-liftable modules. We conclude with Chapter 9, which presents a list of open problems.

In the past seven years, there has been a lot of activity on these topics, and about a hundred papers have appeared on the topic. The topics covered in this monograph reflect the personal preferences of authors, and we do not claim to provide an exhaustive survey of the area. We sincerely believe that the tools developed in this monograph will help researchers to employ new techniques in solving various long-standing open problems in the theory of modules.

Pedro A. Guil Asensio would like to dedicate this monograph to his partner Rosa. The work of Pedro A. Guil has been partially supported by the Spanish Government under grant number MTM2016-77445-P, which includes FEDER funds of the EU, and by Fundación Séneca of Murcia under grant number 19880/GERM/15.

Ashish K. Srivastava would like to dedicate this monograph to the loving memory of his mother Giriza Srivastava and his son Anupam Srivastava, a budding astronomer who left Planet Earth too soon. He would like to thank his daughter Ananya and son Arnab for giving him a reason to live and smile, and he would also like to thank his father Suresh C. Srivastava; wife Sweta; siblings Ashutosh and Amitabh; Amrita and other family members and friends for their support and encouragement. The work of Ashish K. Srivastava is partially supported by a grant from Simons Foundation (grant number 426367).

Askar Tuganbaev would like to dedicate this monograph to his wife Natalie, his sister Saule, and his children and grandchildren Saule, Askar, Diar, Timur, Arman, Maxim, Ivar, Polina, Anvar, Dina and Arthur. The work of Askar Tuganbaev is supported by Russian Scientific Foundation, project 16-11-10013P.

1

Preliminaries

In this chapter, we review what will be needed to understand the content of this monograph. Most of the basic facts will be stated without proof.

1.1 Basics of Ring Theory and Module Theory

We begin by recalling basic definitions from ring theory and module theory. An additive abelian group R with addition $+$ is called a ring if R is also a multiplicative monoid with respect to one more operation of multiplication and

$$x(y+z) = xy + xz, \quad (y+z)x = yx + zx \quad \text{for all} \quad x, y, z \in R.$$

All our rings are assumed to contain a nonzero identity element. For a ring R, the center $Z(R)$ is the center of the multiplicative monoid $(R, \cdot)$, i.e. $Z(R) = \{c \in R \mid ca = ac \text{ for all } a \in R\}$, and $Z(R)$ contains the zero element 0_R of the ring R. If R is a ring and there exists a positive integer $n \in \mathbb{N}$ such that $na = 0$ for all $a \in R$, then the least positive integer n with this property is called the characteristic of R; it is denoted by char R. By our definition, any ring R contains at least two elements: the zero element 0_R and the identity element $1_R \neq 0_R$ contained in the center $Z(R)$ of R.

A ring R is called a division ring if every nonzero element of R is invertible (i.e. the set of all nonzero elements of R is a group with respect to multiplication). A commutative division ring is called a field. An element e of the semigroup X is called an idempotent if $e * e = e$. If e is a nonzero idempotent in R, then eRe is a subring of R with identity element e, and eRe is not a unitary subring of R for $e \neq 1$. The idempotents 0 and 1 of the ring R are central in R; they are called trivial idempotents, and the remaining idempotents of R are called nontrivial. If $\{e_i\}_{i \in I}$ is some set of idempotents in the ring R

and $e_i e_j = e_j e_i = 0$ for all $i \neq j$, then e_i are called **orthogonal** idempotents. If $e_1, \ldots, e_n$ are orthogonal idempotents of R and $e_1 + \cdots + e_n = 1_R$, then $\{e_i\}_{i=1}^n$ is called a **complete set of orthogonal idempotents** of R. For any (resp., central) idempotent $e \in R$, the set $\{1, 1 - e\}$ is a complete set of (resp., central) orthogonal idempotents.

Let R be a ring, and let M be an additive abelian group. The group M is called a **right** (resp., **left**) **R-module** if, for any $m \in M$ and $a \in R$, the element ma in M (resp., am in M) is uniquely defined, $(x + y)a = xa + ya$ (resp., $a(x + y) = ax + ay$), $x1 = x$ (resp., $1x = x$), and $x(ab) = (xa)b$ (resp., $(ba)x = b(ax)$) for all $x, y \in M$ and $a, b \in R$. We denote by M_R (resp., $_R M$) the property that M is a right (resp., left) R-module.

If M is a right (resp., left) R-module and X is a subgroup of the additive group M such that $xa \in X$ (resp., $ax \in X$) for all $x \in X$ and $a \in R$, then X is called a **submodule** of M. A submodule X of M is said to be **proper** if $X \neq M$. A subgroup I of the additive group $(R, +)$ of the ring R is called a **right** (resp., **left**) **ideal** of R if $xr \in I$ (resp., $rx \in I$) for all $r \in R$ and $x \in I$. Thus, right (resp., left) ideals of R coincide with submodules of the module R_R (resp., $_R R$). An **ideal** or a **two-sided ideal** of R is a subset of R that is both a left and a right ideal.

Let A, B be two rings, and let M be a left A-module that also is a right B-module and $(am)b = a(mb)$ for all $m \in M$, $a \in A$, and $b \in B$. Then M is called an **A-B-bimodule**. We denote by $_A M_B$ the property that $_A M_B$ is an A-B-bimodule. If X is a subset of $_A M_B$ such that $ax \in X$ and $xb \in X$ for all $x \in X$, $a \in A$, and $b \in B$, then X is called a **sub-bimodule** of $_A M_B$.

If X is a subset of a right (resp., left) R-module M, then we denote by XR (resp., RX) the submodule of M consisting of all finite sums $\sum x_i a_i$ (resp., $\sum a_i x_i$) in X, where $x_i \in X$ and $a_i \in R$. Thus, for any subset X of the right R-module (resp., left R-module) M, we have the submodule XR (resp., RX); if X consists of a single element x, this submodule is called a **cyclic module** with **generator** x. Thus, for any subset B of R, we have the right ideal BR and the left ideal RB; if B consists of a single element b, this right (resp., left) ideal is called a **principal** right (resp., left) ideal with **generator** x.

The ring R is an (R_1, R_2)-bimodule for any unitary subrings R_1, R_2 of R. In particular, the ring R is an (R, R)-bimodule, and its sub-bimodules coincide with ideals of R. Every additive abelian group M is turned into a module over the ring of integers $\mathbb{Z}$ if for any $m \in M$ and $n \in \mathbb{N}$, we assume that

$$xn = \underbrace{x + \cdots + x}_{n \text{ times}}, \quad x(-n) = -xn, \quad x0_{\mathbb{Z}} = 0_M.$$

Thus, abelian groups coincide with $\mathbb{Z}$-modules.

Let X and Y be two additive abelian groups. A mapping $f: X \to Y$ is called a **group homomorphism** if $f(x + y) = f(x) + f(y)$, $f(-x) = -f(x)$ and $f(0_X) = 0_Y$ for all $x, y \in X$. The subset $\{f(x) \mid x \in X\}$ in Y is called the **image** of the homomorphism f; it is denoted by $f(X)$ or $\mathrm{Im}(f)$. The subset $\{x \in X \mid f(x) = 0\}$ in X is called the **kernel** of the homomorphism f; it is denoted by $\mathrm{Ker}\, f$. If $f(X) = Y$ (resp., $\mathrm{Ker}\, f = 0$), then f is called a **surjective** homomorphism (resp., an **injective** homomorphism or a **monomorphism**). If f is a surjective monomorphism, then f is called an **isomorphism**. We denote by $\mathrm{Hom}_{\mathbb{Z}}(X, Y)$ the set of homomorphisms from X into Y; it is an additive group with addition defined by the relation $(f + g)(x) = f(x) + g(x)$ for all $x \in X$.

Let X_1 be a subgroup in the abelian group X. For every $x \in X$, the set $\{x + x_1 \mid x_1 \in X_1\}$ is denoted by $x + X_1$. We denote by X/X_1 the set $\{x + X_1 \mid x \in X\}$. For any two elements $x, y \in X$, we denote by $(x + X_1) + (y + X_1)$ the subset $(x + y) + X_1$ in X/X_1. It may be easily checked that this addition turns X/X_1 into an additive abelian group with zero element $0 + X_1 = X_1$. The group X/X_1 is called the **factor group** of the group X with respect to the subgroup X_1. The relation $x \to x + X_1$ defines a surjective homomorphism $h: X \to X/X_1$ which is called the **natural epimorphism**. Every surjective group homomorphism $f: X \to Y$ induces the group isomorphism $\overline{f}: X/\mathrm{Ker}\, f \to Y$ which is defined by the relation $\overline{f}(x + \mathrm{Ker}\, f) = f(x)$.

Let R be a ring, let X be a right (resp., left) R-module, let X_1 be a submodule in X, and let X/X_1 be the factor group of the additive group X with respect to Y. For any two elements $x \in X$ and $a \in R$, we set $(x + X_1)a = xa + X_1$ (resp., $a(x + X_1) = ax + X_1$). It is directly verified that X/X_1 is a right (resp., left) R-module; X/X_1 is called the **factor module** of X with respect to X_1. For a module M, a submodule of any factor module of M is called a **subfactor** of M. If R is a ring and I is an ideal of R, then it is directly verified that the factor module R_R/I is turned into a ring, in which multiplication is defined by the relation $(a_1 + I)(a_2 + I) = a_1 a_2 + I$ for all $a_1, a_2 \in R$. The ring R/I is called the **factor ring** of the ring R with respect to the ideal I.

If X and Y are right (resp., left) R-modules and $f: X \to Y$ is a homomorphism of additive groups such that $f(xa) = f(x)a$ (resp., $(ax)f = a(x)f$) for all $x \in X$ and $a \in R$, then f is called an **R-module homomorphism**. If $f: X \to Y$ is a module homomorphism and Y' is a submodule of Y, then we denote by $f^{-1}(Y')$ the submodule $\{x \in X \mid f(x) \in Y'\}$ of X. We assume that homomorphisms of right (resp., left) modules act on the elements from the left (resp., from the right). In addition, if $f: X \to Y$ and $g: Y \to Z$

are two homomorphisms of right (resp., left) modules, then the composition $gf: X \rightarrow Z$ (resp., $fg: X \rightarrow Z$ of the homomorphisms f and g is defined by the relation $gf(x) = g(f(x))$ (resp., $(x)fg = ((x)f)g$).

Let A, B be two rings, let X and Y be two A-B-bimodules, and let $f: X \rightarrow Y$ be a homomorphism of left A-modules that also is a homomorphism of right B-modules. Then f is called a homomorphism of A-B-bimodules. Isomorphisms, endomorphisms and automorphisms of A-B-bimodules are similarly defined. Let X and Y be two rings. A mapping $f: X \rightarrow Y$ is called a ring homomorphism if f is a homomorphism of additive groups and a homomorphism of multiplicative monoids. A (ring or module) surjective monomorphism is called an isomorphism.

Let R be a commutative ring, and let A be a ring. The ring A is called an algebra over R or an R-algebra if there exists a nonzero ring homomorphism f from R into the center of the ring A. Then $f(R)$ is a central unitary subring in A, and we identify R with $f(R) \subseteq A$ provided that $\text{Ker } f = 0$. Thus, if R is a field, then $\text{Ker } f = 0$, and we can assume that the ring A is an algebra over the field R if and only if A contains the field R as a central unitary subring.

If X is a module (resp., a ring), then module (ring) homomorphisms $X \rightarrow X$ are called module (resp., ring) endomorphisms of the module (resp., ring) X, and for any submodule (resp., ideal) X' of X, by the relation $x \rightarrow x + X'$ is defined the surjective module (resp., ring) homomorphism $h: X \rightarrow X/X'$, called the natural epimorphism. The kernel $\text{Ker } f$ of any module (resp., ring) homomorphism $f: X \rightarrow Y$ is a submodule (resp., an ideal) of the module (resp., of the ring) X, and the relation $\overline{f}(x + \text{Ker } f) = f(x)$ induces the module (ring) isomorphism $\overline{f}: X/\text{Ker } f \rightarrow f(X)$. If X is a module (resp., a ring) and Y is a submodule (resp., an ideal) of X, then every submodule (resp., every ideal) of the factor module (resp., of the factor ring) X/Y has the form Y'/Y, where Y' is a submodule (resp., an ideal) of X and $Y \subseteq Y' \subseteq X$, and the relation $\varphi(x + Y') = (x + Y) + Y'/Y$ defines a module (resp., ring) isomorphism $X/Y' \cong (X/Y)/(Y'/Y)$.

For any two right (resp., left) R-modules X and Y, the set of all homomorphisms from X into Y is denoted by $\text{Hom}(X_R, Y_R)$ (resp., $\text{Hom}(_RX, _RY)$); it is a subgroup of the additive group $\text{Hom}_{\mathbb{Z}}(X, Y)$. The additive group $\text{Hom}(X_R, X_R)$ (resp., $\text{Hom}(_RX, _RX)$) is denoted by $\text{End } X_R$ (resp., $\text{End}_R X$); it is a ring such that the product fg of two endomorphisms f and g coincides with the composition of f and g, i.e. $fg(x) = f(g(x))$ (resp., $(x)fg = ((x)f)g$). The ring $\text{End } X_R$ (resp., $\text{End}_R X$) is called the endomorphism ring of the right module X_R (resp., the left module $_RX$). Every ring R is isomorphic to the endomorphism rings $\text{End } R_R$ and $\text{End}_R R$; the required ring isomorphisms are the mappings $\varphi \colon R \to \operatorname{End} R_R$ and $\psi \colon R \to \operatorname{End}_R R$ such that $\varphi(a) \colon x \to ax$ and $\psi(a) \colon x \to xa$ for all $x \in R$.

A module (resp., ring) endomorphism is called a module (resp., ring) auto-morphism if it is both injective and surjective. The set of all automorphisms of the module (ring) X is denoted by $\operatorname{Aut}(X)$. It is a group such that for any auto-morphism f, the inverse automorphism f^{-1} is correctly defined by the relation $f^{-1}(f(x)) = x$ for all $x \in X$. The identity element of the automorphism group is the identity automorphism $1_X \colon x \to x$. The automorphism group of any module is the group of invertible elements of the endomorphism ring of this module. Injective homomorphisms are also called monomorphisms, and surjective homomorphisms are called epimorphisms.[1]

If X and Y are two right R-modules, then for any $f \in \operatorname{End} Y_R$, $g \in \operatorname{Hom}(X_R, Y_R)$ and $h \in \operatorname{End} X_R$, the composition fgh is contained in the group $\operatorname{Hom}(X_R, Y_R)$. Similarly, if X and Y are left R-modules, then for any $f \in \operatorname{End}_R Y$, $g \in \operatorname{Hom}(_R X, _R Y)$ and $h \in \operatorname{End}_R X$, the composition hgf is contained in the group $\operatorname{Hom}(_R X, _R Y)$. Therefore, there are natural bimodules $_{\operatorname{End} Y_R} \operatorname{Hom}(X_R, Y_R)_{\operatorname{End} X_R}$, and similarly, we have $_{\operatorname{End}_R X} \operatorname{Hom}(_R X, _R Y)_{\operatorname{End}_R Y}$.

For a module M, a submodule X of M is said to be fully invariant in M if $f(X) \subseteq X$ for any endomorphism f of M. Every fully invariant submodule of the right module M also is a submodule of the left module $_{\operatorname{End} M} M$. For a ring R, the ideals of R coincide with the fully invariant submodules in R_R; they also coincide with the fully invariant submodules in $_R R$. If a left ideal X of R is not an ideal in R, then X is a submodule of the left $\operatorname{End} R_R$-module R, and X is not a fully invariant submodule of R_R.

Let R be a ring. If X, Y are two subsets of a right (resp., left) R-module M, then we set
$(X \cdot Y) = \{a \in R \colon Xa \subseteq Y\}$ and $r_R(X) = r(X) = (X \cdot 0) = \{a \in R \colon Xa = 0\}$
(resp., $(Y \cdot X) = \{a \in R \colon aX \subseteq Y\}$ and $\ell_R(X) = \ell(X) = (0 \cdot X)) = \{a \in R :: aX = 0\}$).

The subset $r(X)$ (resp., $\ell(X)$) of the ring R is a right (left) ideal of R; it is called the right (resp., left) annihilator of the subset X of the right (resp., left) module M.

If Y is a submodule of a right (resp., left) module M, then $(X \cdot Y)$ (resp., $(Y \cdot X)$) is a right (resp., left) ideal of R, and $(X \cdot Y)$ (resp., $(Y \cdot X)$) is an ideal of R, provided that X and Y are submodules of the right (left) module M.

---

[1] Here we do not consider epimorphisms and monomorphisms in the categorical sense. We only note that the category of rings has nonsurjective epimorphisms, but natural ring epimorphisms onto factor rings are surjective, and all epimorphisms (resp., monomorphisms) in the category of right R-modules are surjective (resp., injective).

It is clear that $r(X) = \cap_{x \in X} r(x)$ (resp., $\ell(X) = \cap_{x \in X} \ell(x)$) and $r(X)$ (resp., $\ell(X)$) is an ideal of R if X is a submodule of the right (resp., left) module M.

A module with zero annihilator is called a faithful module. For a ring R, every right R-module M can be naturally turned into a faithful right module over the ring $R/r(M)$. In addition, if we associate with any element $m \in M$ the homomorphism $f_m \in \mathrm{Hom}(R_R, M)$ such that $f_m(a) = ma$ for all $a \in R$, then we have an End M-R-bimodule isomorphism $M \to \mathrm{Hom}(R_R, M)$.

For any cyclic module xR, the mapping $f : a \to xa$ is a module epimorphism from R_R onto xR with kernel $r(x)$; it induces the module isomorphism $R_R/r(x) \cong xR$. Similarly, $Rx \cong_R R/\ell(x)$. This implies that all cyclic right (resp., left) R-modules coincide, up to isomorphism, with the factor modules R/B with respect to right (resp., left) ideals B and cardinalities of all cyclic R-modules are upper bounded by the cardinality of R. Therefore, there exists a set $\mathcal{E}$ of cyclic right R-modules such that any cyclic right R-module is isomorphic to some module in $\mathcal{E}$.[2]

If R is a ring and X is a subset in R, then $r(\ell(r(X))) = r(X)$ and $\ell(r(\ell(X))) = \ell(X)$. If Q is a ring and R is a subring in Q, then for every subset X in R, we have that $r_R(X) = R \cap r_Q(X)$ and $\ell_R(X) = R \cap \ell_Q(X)$.

Theorem 1.1 *For a ring R, the following conditions are equivalent.*

1. *R is a ring with maximum condition on right annihilators.*
2. *R is a ring with minimum condition on left annihilators.*
3. *R is a subring of a ring with maximum condition on right annihilators.*
4. *R is a subring of a ring with minimum condition on left annihilators.*

In this case, any subset B in R contains a finite subset $B^* = \{b_1, \ldots, b_n\}$ such that

$$r(B) = r(B^*) = r(b_1) \cap \cdots \cap r(b_n).$$

Theorem 1.2 *For a ring R, the following conditions are equivalent.*

1. *For every subset $X \subseteq R$, there exists an idempotent $e \in R$ with $r(X) = eR$.*
2. *For every subset $Y \subseteq R$, there exists an idempotent $f \in R$ with $\ell(Y) = Rf$.*

---

[2] The similar assertion does not hold for all right R-modules, since there does not exist a common upper bound of cardinalities of all right R-modules.

Proof It is sufficient to prove the implication $(1) \Rightarrow (2)$. We set $X = \ell(Y)$. We have $X = \ell(r(X))$. By assumption, $r(X) = eR$ for some idempotent $e \in R$. Then $Y = \ell(eR) = R(1 - e)$, and we set $f = 1 - e$.   □

A ring R is called a Baer ring if it satisfies the preceding equivalent conditions (1) and (2).

1.2 Simple and Semisimple Modules

A nonzero module M is said to be a simple module if it has no nonzero proper submodule – that is, the only submodules of M are zero module and the module M itself. A ring R is said to be a simple ring if R has no nonzero proper two-sided ideal.

Let M a nonzero module. A submodule N of M is said to be a maximal submodule if N is a maximal element of the nonempty set of all proper submodules of M with respect to set-inclusion. The set of all maximal submodules of M is denoted by max M.

For a module M, the intersection of kernels of all homomorphisms from M into arbitrary simple modules is called the Jacobson radical of M; it is denoted by $J(M)$. It is clear that either $J(M) = M$ (if max $M = \varnothing$) or $J(M)$ coincides with the intersection of all maximal submodules of M (if max $M \neq \varnothing$). A module M is said to be a semiprimitive module if $J(M) = 0$.

For a module M, a submodule X of M is said to be small (in M) if $X + Y \neq M$ for any proper submodule Y of M. If X is a small submodule of M, then we express it as $X \subset_s M$. Every nonzero submodule has a simple subfactor.

Theorem 1.3 *For a nonzero module M, the following conditions are equivalent.*

1. *M is the direct sum of simple modules.*
2. *M is the sum of simple modules.*
3. *Every submodule of M is a direct summand of M.*

A module M is said to be semisimple if M satisfies the preceding equivalent conditions.

Theorem 1.4 (Wedderburn–Artin theorem) *For a ring R, the following conditions are equivalent.*

1. *R is a right (resp., left) semisimple ring.*
2. *Each right (resp., left) R-module is semisimple.*
3. *R is a finite direct product of simple Artinian rings.*
4. *R is isomorphic to a finite direct product of matrix rings over division
   rings.*

For a module M, the sum of all simple submodules of M is called the socle of
M; it is denoted by $\mathrm{Soc}(M)$. If M has no simple submodules, then $\mathrm{Soc}(M) = 0$
by definition. Clearly, $\mathrm{Soc}(M)$ is a fully invariant submodule of M, and it is
the largest semisimple submodule of M.

Definition 1.5 A right R-module M is called a semi-Artinian module if for
every submodule $N \neq M$, $\mathrm{Soc}(M/N) \neq 0$. A ring R is called a right semi-
Artinian ring if R_R is semi-Artinian.

If X is a module (resp., a ring) and $\{Y_i\}_{i \in I}$ is a set of submodules (resp.,
ideals) of X with $\cap_{i \in I} Y_i = 0$, then X is called a subdirect product of
the factor modules X/Y_i (resp., factor rings X/Y_i). In this case, if at least
one of the modules (resp., ideals) Y_i is equal to zero, then the subdirect
product is said to be trivial. If $Y_i \neq 0$ for all i, then the subdirect product
is said to be nontrivial. A nonzero module (resp., ring) X is said to be
subdirectly indecomposable if X is not a nontrivial subdirect product
of any factor modules (resp., factor rings) of X; i.e. the intersection of
all nonzero submodules (resp., ideals) of X does not equal to zero. This
means that there exists a nonzero submodule (resp., ideal) of X contained
in every nonzero submodule (resp., ideal) of X. Every nonzero module
(resp., ring) is a subdirect product of subdirectly indecomposable modules
(resp., rings).

For a module M, the Jacobson radical $J(M)$ is fully invariant in M,
$M/J(M)$ is a semiprimitive module and $M/J(M)$ is a subdirect product of
simple modules. The Jacobson radical $J(M)$ contains each small submodule
X of M and every finitely generated submodule $N \subseteq J(M)$ is small in M.
Therefore, $J(M)$ is the sum of all small submodules of M. If $J(M)$ is small
in M, then $J(M)$ is the largest small submodule of M. In addition, if M is a
nonzero finitely generated module, then $M \neq J(M)$ and $J(M)$ is the largest
small submodule of M.

For any ring R, we have $J(R_R) = J(_R R)$; this ideal is denoted by $J(R)$.
The ideal $J(R)$ is called the Jacobson radical of the ring R. It coincides with
the largest ideal I in R with the property that $1 - x$ is invertible in R for all
$x \in I$.

1.3 Essential and Closed Submodules

Let M be a module. If X is a submodule of M such that $X \cap Y \neq 0$ for any nonzero submodule Y of M, then X is said to be essential in M. In this case, we say that M is an essential extension of X, and we express it as $X \subseteq_e M$. A submodule Y of M is said to be closed (in M) if Y coincides with any submodule in M that is an essential extension of Y. If $X, \overline{X}$ are submodules of M and $\overline{X}$ is a closed (in M) essential extension of X, then $\overline{X}$ is called the closure of X in M.

Let X be a submodule of M, and let Y be a closed submodule of M such that $X \cap Y = 0$, M is an essential extension of $X \oplus Y$, and $X \cap Y' \neq 0$ for any submodule Y' of M properly containing Y. Then Y is called a ∩-complement to X in M. A submodule Y of M is said to be ∩-complement if Y is a ∩-complement in M to some submodule of M.

We list some useful facts in the next theorem.

Theorem 1.6 *Let M be a nonzero module over a ring R.*

1. *If M is an essential extension of a module X, then M is an essential extension of any essential submodule Y of X, X contains the socle of M, and N is an essential extension of $X \cap N$ for any submodule N of M.*
2. *Every direct summand of M is closed in M, and every closed submodule X of M coincides with the closure of X.*
3. *If M is an essential extension of some module N, then for every module homomorphism $f : X \to M$, the submodule $f^{-1}(N)$ is essential in X, where $f^{-1}(N) = \{x \in X : f(x) \in N\}$.*
4. *If $M = \oplus_{i \in I} M_i$ and each M_i is an essential extension of some module X_i, $i \in I$ then M is an essential extension of $\oplus_{i \in I} X_i$.*
5. *M is an essential extension of the direct sum of its nonzero cyclic submodules.*
6. *Every submodule X of M has at least one closure $\overline{X}$, and $\overline{X} \cap Y = 0$ for any submodule Y of M with $X \cap Y = 0$.*
7. *If the sum $\sum_{i \in I} X_i$ of submodules X_i of M is a direct sum, then the sum of closures $\sum_{i \in I} \overline{X}_i$ of submodules X_i in M is a direct sum.*
8. *If X is a submodule of M, then for any submodule Y of M with $X \cap Y = 0$ (e.g. for $Y = 0$), the module M contains at least one closed ∩-complement $\overline{Y}$ to X with $\overline{Y} \supseteq Y$. Therefore, X has at least one closed ∩-complement Y and at least one closure $\overline{X}$ such $\overline{X} \cap Y = 0$, and X is a direct summand of the essential submodule $X' = X \oplus Y$ in M. In particular, X is a direct summand of some essential submodule of M.*

9. *Let X be a submodule of M, let Z be a closure of X in M, and let Y be a
   ∩-complement to X in M. Then Z is a ∩-complement to Y in M, and Y is
   closed in M.*
10. *The set of all closed submodules of M coincides with the set of all
    ∩-complement submodules of M.*
11. *For any module N and each homomorphism $f : X \to N$, there exists an
    essential submodule X' of M and a homomorphism $f' : X' \to N$ such
    that f' coincides with f on X, $f'(X') = f(X)$, $X' = X \oplus Y$, Y is a
    closed ∩-complement to X in M and $f'(Y) = 0$.*

Definition 1.7 A nonzero module M is called a uniform module if any two
nonzero (cyclic) submodules of M intersect nontrivially.

It is clear that every subdirectly indecomposable module is uniform.

Theorem 1.8 *For the module M, the following conditions are equivalent.*

1. *M is a uniform module.*
2. *M is an essential extension of a uniform module.*
3. *M is an essential extension of any nonzero (cyclic) submodule of M.*
4. *Any two closed nonzero submodules of M have the nonzero intersection.*

Definition 1.9 A module M is said to be finite-dimensional if M does not
contain an infinite direct sum of nonzero submodules.

All uniform modules are finite-dimensional.

Theorem 1.10 *For the module M, the following conditions are equivalent.*

1. *M is a finite-dimensional module.*
2. *M is an essential extension of a finite direct sum of finite-dimensional
   modules.*
3. *M does not contain an infinite direct sum of closed submodules.*
4. *M is a module with maximum condition on closed submodules.*
5. *M is a module with minimum condition on closed submodules.*
6. *Any submodule of M is an essential extension of a finitely generated
   module.*
7. *There exists a finite set $\{X_i\}_{i=1}^{k}$ of submodules X_i of M such that
   $\cap_{i=1}^{k} X_i = 0$ and M/X_i is a uniform module, $i = 1, \ldots, k$.*
8. *There exists a positive integer n such that M does not contain a direct sum
   of $n + 1$ nonzero modules, and M is an essential extension of the direct
   sum of n nonzero uniform modules.*

The integer n from condition (6) is called the uniform dimension or
the Goldie dimension of the finite-dimensional module M; it is denoted

by u. dim M. If M is a module of finite uniform dimension n, then every nonzero submodule X of M is the direct sum of at most n nonzero indecomposable modules, and every set of nonzero orthogonal idempotents of the ring End X contains at most n elements. If $M = \oplus_{i=1}^{m} X_i = \oplus_{j=1}^{n} Y_j$, where all X_i and Y_j are nonzero uniform modules, then $m = n$.

A family of subsets $\{S_i : i \in \mathcal{I}\}$ in a set S is said to satisfy the *Ascending Chain Condition* (*ACC* in short) if for any ascending chain $S_{i_1} \subseteq S_{i_2} \subseteq \cdots$ in the family, there exists an integer n such that $S_{i_n} = S_{i_{n+k}}$ for each $k \in \mathbb{N}$.

A family of subsets $\{S_i : i \in \mathcal{I}\}$ in a set S is said to satisfy the *Descending Chain Condition* (*DCC* in short) if for any descending chain $S_{i_1} \supseteq S_{i_2} \supseteq \cdots$ in the family, there exists an integer n such that $S_{i_n} = S_{i_{n+k}}$ for each $k \in \mathbb{N}$.

Definition 1.11 A module M is called a Noetherian module (resp., an Artinian module) if M does not contain an infinite properly ascending (resp., descending) chain of submodules.

A ring R is called *right Artinian* (resp., *Noetherian*) if R_R is Artinian (resp., Noetherian).

Theorem 1.12 (Hopkins–Levitzki theorem) *If R is a right Artinian ring, then R is also right Noetherian and $J(R)$ is nilpotent.*

This was proved by Hopkins and Levitzki. Note that an Artinian module need not be Noetherian. The ring of p-adic integers as a module over $\mathbb{Z}$ is Artinian but not Noetherian.

Definition 1.13 A module M is said to be locally Noetherian (resp., locally Artinian) if every cyclic submodule of M is Noetherian (resp., Artinian). A module M is said to be quotient finite-dimensional if all factor modules of M are finite-dimensional.

Theorem 1.14 *Let M be a right R-module.*

1. *If R is a right Noetherian (resp., right Artinian) ring, then M is a locally Noetherian (resp., locally Artinian) module.*
2. *If M is a Noetherian or Artinian module, then M is quotient finite-dimensional.*
3. *The module M is quotient finite-dimensional if and only if M does not have a subfactor that is an infinite direct sum of simple modules.*

Definition 1.15 Let R be a ring and M be a right R-module. We denote by $Z(M)$ the set of all $m \in M$ such that $r(m)$ is an essential right ideal of R. The module M is called a singular module if $Z(M) = M$. The module M is said to be a nonsingular module if $Z(M) = 0$.

Note that $Z(R_R)$ and $Z(_R R)$ are two-sided ideals of R that are called the right singular ideal and the left singular ideal, respectively.

We denote by $\Delta(M)$ the set of all endomorphisms of M whose kernels are essential submodules of M. We have $f(Z(M)) \subseteq Z(M)$ for every homomorphism $f: Z(M) \to M$. In particular, $Z(M)$ is a fully invariant singular submodule of M; it is called the singular submodule of M. It may be noted that $Z(M)$ contains all singular submodules of M, and $g(Z(M)) \subseteq Z(Y)$ for every module homomorphism $g: M \to M'$. In addition, $Z(X) = 0$ for any submodule X of M with $X \cap Z(M) = 0$.

We denote by $G(M)$ the intersection of all submodules X of the module M such that the factor-module M/X is nonsingular; it is also denoted by $Z_2(M)$ in the literature. The submodule $G(M)$ is called the Goldie radical or the second singular submodule of the module M. Furthermore, $f(G(M)) \subseteq G(M)$ for every homomorphism $f: G(M) \to M$. In particular, $G(M)$ is a fully invariant submodule of M, and the right Goldie radical $G(R_R)$ and the left Goldie radical $G(_R R)$ of the ring R are ideals of R.

We list some useful facts in the following theorem.

Theorem 1.16 *Let M be a right R-module.*

1. *The Goldie radical of M, $G(M)$ is a closed submodule in M; it is the unique closure of the singular submodule $Z(M)$ in M.*
2. *If X is a submodule of M and $X \cap Z(M) = 0$, then $X \cap G(M) = 0$ and the module X is nonsingular.*
3. *If X is an essential submodule of M, then the module M/X is singular. Thus, for every $m \in M$, the right ideal $\{a \in R \mid ma \in X\}$ of the ring R is essential.*
4. *If R is a right nonsingular ring, then $G(M) = Z(M)$ and the module $M/Z(M)$ is nonsingular.*
5. *If M is a nonsingular module and X is a closed submodule in M, then M/X is a nonsingular module.*
6. *$\Delta(M)$ is an ideal of the endomorphism ring $\mathrm{End}(M)$, and consequently, we have that $\sum_{h \in \Delta(M)} h(M)$ is a fully invariant submodule of the module M; this submodule is contained in $Z(M)$. In particular, if M is a nonsingular module, then $\Delta(M) = 0$. In addition, $Z(R_R) = \Delta(R_R)$, and $Z(_R R) = \Delta(_R R)$.*
7. *If M is a nonsingular module, then all submodules, essential extensions and subdirect products of M are nonsingular.*
8. *If X is a submodule of M and the module M/X is nonsingular, then X is a closed submodule of M.*

9. *If $\{X_i\}_{i \in I}$ is some set of submodules of M and all modules M/X_i are nonsingular, then M/X is a nonsingular module and $X = \cap_{i \in I} X_i$ is a closed submodule of M. In particular, the factor-module $M/G(M)$ is nonsingular.*
10. *If $f : M' \to M$ is a module homomorphism such that Ker f is an essential submodule, then $f(M')$ is a singular module which is contained in $Z(M)$.*
11. *If M is not singular, then M contains an isomorphic copy of a nonzero right ideal of R. In addition, if R is right finite-dimensional, then M contains an isomorphic copy of a nonzero uniform right ideal of R.*

Theorem 1.17 *Every Baer ring is (right and left) nonsingular.*

1.4 Prime Rings and Semiprime Rings

Definition 1.18 A ring R is said to be a prime ring if the product of any two nonzero ideals of R does not equal to zero.

A right or left ideal P in R is said to be prime if $xRy \not\subseteq P$ for any $x, y \in R \setminus P$. It is not difficult to see that a proper ideal P in R is prime if and only if the factor ring R/P is prime. It follows that the ring R is prime if and only if the product XY of any two nonzero right ideals X and Y of R does not equal to zero. This follows from the relation $(RXR)(RYR) = RXYR$.

Definition 1.19 A subset $X \subseteq R$ is said to be nilpotent if $X^n = 0$ for some $n \in \mathbb{N}$ – i.e. $x_1, \ldots, x_n = 0$ for any elements $x_1, \ldots, x_n \in X$. The least such integer n is called the nilpotence index of X.

Definition 1.20 A ring R is said to be a semiprime ring if R does not have nonzero nilpotent ideals.

It turns out that a ring R is semiprime if and only if $X^2 \neq 0$ for any nonzero right ideal X in R. This follows easily in view of the relation $(RXR)^2 = RX^2R$.

Definition 1.21 An element $a \in R$ is said to have index of nilpotence at most n if $a^n = 0$ for some $n \in \mathbb{N}$.

Definition 1.22 A ring R is called a reduced ring if R does not have nonzero nilpotent elements.

Definition 1.23 A ring R whose all idempotents are central is called an abelian ring.

Every reduced ring R is a nonsingular abelian ring in which the left annihilator of any subset X coincides with the right annihilator of the set X and is an ideal of the ring R.

Definition 1.24 A ring R is called a domain if $ab \neq 0$ for any two nonzero elements $a, b \in R$.

Definition 1.25 A right (resp., left) ideal I of R is called a right (resp., left) nil-ideal if all elements of I are nilpotent.

1.5 Classical Rings of Fractions and Semiprime Goldie Rings

Let R be a ring, and let R^* be the multiplicative monoid of all nonzero-divisors of the ring R.

Theorem 1.26 *Let S be a submonoid of the monoid R^*. The following conditions are equivalent.*

1. *For any $a \in R$ and $s \in S$, there exist elements $b \in R$ and $t \in S$ such that $at = sb$ (resp., $ta = bs$).*
2. *There exists a ring RS^{-1} (resp., a ring $S^{-1}R$) containing R as a unitary subring such that all elements of S are invertible in RS^{-1} (resp., in $S^{-1}R$) and $RS^{-1} = \{as^{-1} \mid a \in R, s \in S\}$ (resp., $S^{-1}R = \{s^{-1}a \mid a \in R, s \in S\}$).*

Under the preceding equivalent conditions, the sumonoid S of R^* is called a right (resp., left) Ore set (in R), and the ring RS^{-1} (resp., $S^{-1}R$) is called the right (resp. left) the ring of fractions of the ring R with respect to S; it is unique up to ring isomorphism.

A submonoid S of the monoid R^* is a right and left Ore set in R if and only if there exists a ring $S^{-1}RS^{-1}$ such that $S^{-1}RS^{-1} = RS^{-1} = S^{-1}R$. In this case, the ring $S^{-1}RS^{-1}$ is unique up to ring isomorphism; it is called the two-sided ring of fractions of the ring R with respect to S.

If $R^* = S$ is a right (resp., left) Ore set in R, the ring RS^{-1} (resp., the ring $S^{-1}R$) is called the right (resp., left) classical ring of fractions of the ring RS^{-1}; it is denoted by $Q_{cl}(R)$ (resp., $_{cl}Q(R)$). In this case, the ring R is called a right (resp., left) order in the ring $Q_{cl}(R)$ (resp., $_{cl}Q(R)$).

In the case, where the monoid $R^* = S$ is a right and left Ore set, then the ring $S^{-1}RS^{-1}$ is called the classical ring of fractions of the ring RS^{-1}; it is denoted by $_{cl}Q_{cl}(R)$. The ring R is called an order in the ring $_{cl}Q_{cl}(R)$.

A domain R is called a right Ore domain if R^*S is a right Ore set – i.e. for any two nonzero elements $a, s \in R$ – there exist elements $b, t \in R$ with $at = sb \neq 0$. In this case, there exists a division ring Q such that R is a unitary subring in Q, every nonzero element $a \in R$ is invertible in Q, and for every nonzero element $q \in Q$, there exist nonzero elements $a, b \in R$ with $q = ab^{-1}$. In this case, the division ring Q is called the classical right division ring of fractions of the domain R, the module Q_R is the injective envelope of the module R_R, and the division ring Q can be naturally identified with the endomorphism ring $\operatorname{End}(Q_R)$.

Right finite-dimensional domains coincide with right Ore domains. In particular, every right Noetherian domain is a right Ore domain.

Theorem 1.27 (Goldie's theorem) *For any ring R, the following conditions are equivalent.*

1. *R is a right order in a semisimple Artinian ring.*
2. *In R, the set of all essential right ideals coincides with the set of all right ideal containing at least one nonzero-divisor.*
3. *R is a right finite-dimensional semiprime ring with the maximality condition on left annihilators.*
4. *R is a semiprime right Goldie ring.*
5. *R is a right finite-dimensional, right nonsingular semiprime ring.*

In this case, R is a ring with the minimality condition on both right annihilators and left annihilators.

Theorem 1.28 *For a ring R, the following conditions are equivalent.*

1. *R is a right order in the matrix ring over division ring.*
2. *R is a right finite-dimensional prime ring with the maximality condition on left annihilators.*
3. *R is a prime right Goldie ring.*
4. *R is a right finite-dimensional right nonsingular prime ring.*

Theorem 1.29 *For a ring R, the following conditions are equivalent.*

1. *R is an order in the matrix ring over a division ring (resp., in the direct product of finite number of matrix rings over division rings).*
2. *R is a prime (resp., semiprime) Goldie ring.*
3. *R is a finite-dimensional nonsingular prime (resp., semiprime) ring.*

Let R be a semiprime right Goldie ring, and let M be a right R-module. We denote by $t(M)$ the set of all elements from M that are annihilated by some nonzero-divisors of the ring R. The module M is called a non-torsion (resp., torsion module; torsion-free module) if $t(M) \neq M$ (resp., $t(M) = M$, $t(M) = 0$).

Theorem 1.30 *Let M be a right R-module.*

1. *M is nonsingular (resp., singular) if and only if M is a torsion module (resp., a torsion-free module).*
2. *If M is a singular module, then any essential extension of M is a singular module. Therefore, $Z(M) = G(M)$.*
3. *If there exists a submodule X of M such that the modules X and M/X are singular, then the module M is singular.*
4. *$Z(M)$ is a closed submodule of M, and the module $M/Z(M)$ is nonsingular.*

1.6 Local, Semilocal and Semiperfect Rings

Definition 1.31 A module M is said to be a local module if M satisfies the following equivalent conditions.

1. M is cyclic, and $M/J(M)$ is a simple module.
2. M is a nonzero cyclic module with unique maximal submodule.
3. $M \neq J(M)$ and $M = xR$ for any $x \in M \backslash J(M)$.

Definition 1.32 A ring R is said to be a local ring if R satisfies the following equivalent conditions.

1. For every element $a \in R$, at least one of the elements $a, 1-a$ is invertible.
2. $R/J(R)$ is a division ring.
3. R_R is a local module.
4. $_R R$ is a local module.
5. $J(R)$ coincides with the set of all noninvertible elements of the ring R.
6. Every cyclic right or left R-modules is local.

Definition 1.33 A nonzero idempotent e of the ring R is said to be a local idempotent if e satisfies the following equivalent conditions.

1. eRe is a local ring.
2. eR_R is a local module.
3. $_R Re$ is a local module.

Definition 1.34 A ring R is called a semilocal ring if $R/J(R)$ is semisimple Artinian.

Definition 1.35 A ring R is called a semiprimary ring if $R/J(R)$ is semisimple Artinian and $J(R)$ is nilpotent.

Definition 1.36 A ring R is said to be a semiperfect ring if R satisfies the following equivalent conditions.

1. R_R is a finite direct sum of local modules.
2. $_R R$ is a finite direct sum of local modules.
3. The identity element of the ring R is the sum of local orthogonal idempotents.
4. The factor ring $R/J(R)$ is semisimple Artinian and idempotents lift modulo $J(R)$.

Definition 1.37 A module M is called a Bezout module if M satisfies the following equivalent conditions.

1. Every finitely generated submodule of M is cyclic.
2. Every two-generated submodule of M is cyclic.
3. For any two elements $x, y \in M$, there exist elements $a, b, c, d \in R$ such that $(xa + yb)c = x$ and $(xa + yb)d = y$.

Definition 1.38 A module M is said to be a uniserial module if M satisfies the following equivalent conditions.

1. Any two submodules of M are comparable with respect to inclusion.
2. Any two cyclic submodules of M are comparable with respect to inclusion.
3. Every finitely generated nonzero submodule of M is a local module.

Definition 1.39 A module M is called a serial module if M is a direct sum of uniserial modules.

Definition 1.40 A ring R is said to be a right uniserial ring if R satisfies the following equivalent conditions.

1. R_R is a uniserial module.
2. R is a right Bezout local ring.

1.7 Injective and Projective Modules

Let R be a ring, and let M be a right R-module. A right R-module X is called a free cyclic module if there exists an element $x \in X$ such that $X = xR$ and $r(x) = 0$; the element x is called a free generator of X. We remark that X_R is a free cyclic module if and only if $X \cong R_R$.

A module X_R is said to be free if there exists a subset $\{x_i\}_{i \in I} \subseteq X$ such that $X = \oplus_{i \in I} x_i R$ and $r(x_i) = 0$ for all $i \in I$; this subset is called a basis of X, and the cardinality card(I) is called the rank of the module X. We remark that the rank of a free module is not necessarily unique.

M is a free module (of rank $\aleph$) if and only if X is isomorphic to the direct sum of some set I (of cardinality $\aleph$) of isomorphic copies of the free cyclic module R_R. In particular, there exist free modules of any rank $\aleph$.

Every mapping f from the basis $\{x_i\}_{i \in I}$ from the free module X_R into any module M_R can be extended to the homomorphism $g: X \to M$ with the use of the relation $g(\sum x_i a_i) = \sum f(x_i) a_i$. In addition, if $\{f(x_i)\}_{i \in I}$ is a generator system of the module M, then $g: X \to M$ is an epimorphism.

Theorem 1.41 *Every module with generator system of cardinality $\aleph$ is a homomorphic of a free module of rank $\aleph$.*

Proof Let $\{m_i\}_{i \in I}$ be a generator system of cardinality $\aleph$ for the module M_R. We take a free module X_R with basis $\{x_i\}_{i \in I}$ of cardinality $\aleph$. We define a mapping $f: \{x_i\}_{i \in I} \to \{m_i\}_{i \in I}$ such that $f(x_i) = m_i$ for all $i \in I$. Clearly, f can be extended to a module epimorphism $X \to M$. $\qquad\qquad \square$

Definition 1.42 Let R be a ring, and let M, X be two right R-modules. The module M is said to be injective with respect to X or X-injective if for every submodule X_1 in X, each homomorphism $X_1 \to M$ can be extended to a homomorphism $X \to M$.

Definition 1.43 An R-module M is called an injective module if M is injective with respect to each R-module.

Definition 1.44 A module M is called a quasi-injective module if M is injective with respect to M.

Definition 1.45 The module M is said to be projective with respect to X or X-projective if for every epimorphism $h: X \to \overline{X}$ and any homomorphism $\overline{f}: M \to \overline{X}$, there exists a homomorphism $f: M \to X$ with $\overline{f} = hf$.

Definition 1.46 An R-module M is called a projective module if M is projective with respect to each R-module.

Definition 1.47 A module M, which is projective with respect to itself, is called a quasi-projective module or a self-projective module.

It is clear that all projective modules are quasi-projective and a cyclic group of any prime order is a quasi-projective non-projective simple $\mathbb{Z}$-module.

Lemma 1.48 *Every module is projective and injective with respect to each semisimple module.*

The preceding lemma follows easily from the property that each submodule of a semisimple module is a direct summand.

Theorem 1.49 *The class $\mathcal{X}$ of all modules X, such that M is X-injective and contains all submodules, homomorphic images and direct sums of modules from $\mathcal{X}$.*

The class $\mathcal{Y}$ of all modules Y, such that M is Y-projective and contains all homomorphic images, submodules and finite direct sums of modules from $\mathcal{Y}$.

Proof We prove only the first assertion, since the second assertion is proved similarly. If M is X-injective, then it is directly verified that M is injective with respect to any submodule of the module X. We assume that $h \colon X \to \overline{X}$ is an epimorphism, $\overline{Y}$ is a submodule in $\overline{X}$, $\overline{g} \in \mathrm{Hom}(\overline{Y}, M)$. We denote by Y the complete pre-image $\overline{Y}$ in X under the action of h. Let h_Y be the restriction of h to Y. By assumption, the homomorphism $\overline{g} h_Y \colon Y \to M$ can be extended to a homomorphism $f \colon X \to M$. Since $\mathrm{Ker}\, h \subseteq Y$, we have $\mathrm{Ker}\, h \subseteq \mathrm{Ker}\, f$. Therefore, f can be extended to some homomorphism $\overline{f} \colon \overline{X} \to M$, and M is $\overline{X}$-injective.

Let $\{X_i\}_{i \in I}$ be a set of modules such that M is X_i-injective for all $i \in I$. Let $Y = \oplus_{i \in I} X_i$, let Y_1 be a submodule in Y, $f_1 \in \mathrm{Hom}(Y_1, M)$, and let $\mathcal{E}$ be the set of all pairs (L, f_L), where L is a submodule in Y, which contains Y_1, and f_L is a homomorphism from L into M that extends f_1. We define the relation $\leq$ on $\mathcal{E}$ such that $(L, f_L) \leq (Q, f_Q)$ if and only if $L \subseteq Q$ and f_L can be extended to a f_Q. We can verify that $\leq$ is a partial order on $\mathcal{E}$, and every nonempty chain in $\mathcal{E}$ has the upper bound. By the Zorn lemma, $\mathcal{E}$ has a maximal element $(\overline{Y}, \overline{f})$. It is sufficient to prove that $\overline{Y} = Y$; this is equivalent to the relations $X_i \subseteq \overline{Y}$ for all $i \in I$. Since M is X_i-injective, we have that the restriction of homomorphism $\overline{f}$ to $X_i \cap \overline{Y}$ can be extended to some homomorphism $f_i \colon X_i \to M$. Let $u \colon (X_i + \overline{Y}) \to M$ be a homomorphism such that $u(x + y) = f_i(x) + \overline{f}(y)$ for all $x \in X_i$ and $y \in \overline{Y}$. This homomorphism is correctly defined (if $x + y = 0$, then $x = -y \in X_i \cap \overline{Y}$ and $u(x + y) = \overline{f}(-y) + \overline{f}(y) = 0$). By construction, $X_i + \overline{Y} = \overline{Y}$. Therefore, $X_i \subseteq \overline{Y}$, which is required. $\square$

Theorem 1.50 *If X is a right R-module, which contains an isomorphic copy of the module R_R and the module M is injective with respect to module X, then M is an injective module.*

Proof First, note that in this case, M is injective with respect to the module R_R. Now, as every right R-module is a homomorphic image of the direct sum of copies of the module R_R, it follows easily from the preceding theorem that M is injective with respect to each right R-module, and consequently, M is an injective module.   □

As a consequence, we have the following corollary.

Corollary 1.51 *For a ring R, if R_R is quasi-injective, then R_R is an injective, module (such a ring R is called a right self-injective ring).*

Theorem 1.52 *Let M be a module.*

1. *All direct summands and direct products of modules, which are injective with respect to the module M, are M-injective. In particular, all direct summands and direct products of injective modules are injective.*
2. *All direct summands and direct sums of modules, which are projective with respect to the module M, are M-projective. In particular, all direct summands and direct sums of projective modules are projective.*

Proof We prove only the first assertion, since the second assertion is proved similarly. Let N be a right R-module, and let $N = \prod_{i \in I} N_i$. It is clear that the M-injectivity of the module N implies the M-injectivity of all modules N_i. Now we assume that all modules N_i are M-injective. Let X be a submodule in M, $f \in \text{Hom}(X, N)$, and let $\pi_i : N \to N_i$ be natural projections. All homomorphisms $\pi_i f : X \to N_i$ can be extended to homomorphisms $g_i : M \to N_i$ defining a natural extension of $g : M \to N$.   □

Theorem 1.53 *If a module M is injective with respect to a module X and there exists a monomorphism $f : M \to X$, then $f(M)$ is a direct summand of X, M is quasi-injective, and M is isomorphic to a direct summand of the module X. In particular, if either the module X is indecomposable or $f(M)$ is an essential submodule of X, then $f : M \to X$ is an isomorphism.*

Proof M is an $f(M)$-injective module. Since $f(M) \cong M$, we have that M is quasi-injective. Since $f(M)$ is X-injective, we have that natural embedding $f(M) \to X$ can be extended to a homomorphism $g : X \to f(M)$. Then g is the projection from X onto $f(M)$. Therefore, $f(M)$ is a direct summand of X.   □

Theorem 1.54 *If a module M is projective with respect to a module X and there exists an epimorphism $h : X \to M$, then Ker h is a direct summand of X,* *and M is a quasi-projective module that is isomorphic to a direct summand of the module X. In particular, if the module X is indecomposable, then $h: X \to M$ is an isomorphism.*

Proof M is a quasi-projective module. Since M is an X-projective module, there exists a homomorphism $g: M \to X$ with $1_M = hg$. We set $\pi \equiv 1 - gh \in$ End X. Since $\pi^2 = 1 - gh - gh + g(hg)h = 1 - gh = \pi$, we have $X = \pi(X) \oplus (1 - \pi)(X)$. Further,

$$h\pi(X) = (h - hgh)(X) = (h - h)(X) = 0, \quad \pi(X) \subseteq \text{Ker } h,$$

$$\text{Ker } h = gh(\text{Ker } h) + (1 - gh)(\text{Ker } h) = \pi(\text{Ker}(h)) \subseteq \pi(X).$$

Therefore, Ker $h = \pi(X)$ and $M \cong (1 - \pi)(X)$. $\qquad\qquad\qquad\square$

Theorem 1.55 *If Y is a submodule of the module X and the module X/Y is projective with respect to X, then Y is a direct summand in X. In addition, the cyclic module xR is projective if and only if $r(x)$ is a direct summand in R_R if and only is $r(x) = eR$ for some idempotent $e \in R$.*

Proof The first assertion follows from Theorem 1.54. The second assertion follows from the first assertion and the property that for every cyclic module xR, there exists an isomorphism $f: xR \to R_R/r(x)$ such that $f(xa) = a + r(x)$ for all $a \in R$. $\qquad\qquad\qquad\square$

Theorem 1.56 *Every free module is projective.*

Proof Since direct sum of projective modues is projective, it is sufficient to prove that R_R is a projective module. Let $h: X \to \bar{X}$ be an arbitrary epimorphism of right R-modules, and let $\bar{f}: R_R \to \bar{X}$ be a homomorphism. There exists an element $x \in X$ with $h(x) = \bar{f}(1)$. The mapping $f(a) = xa$ is a well-defined homomorphism from R_R into X. In addition, $hf = \bar{f}$. $\qquad\square$

Theorem 1.57 *For a module P, the following statements are equivalent:*

1. *P is projective.*
2. *P is isomorphic to a direct summand of a free module.*
3. *For every module epimorphism $h: X \to P$ where X is any module, the module Ker h is a direct summand of P.*

Theorem 1.58 *(Dual basis lemma) For a right R-module M, the following conditions are equivalent.*

1. *M is a projective module.*
2. *There exist a subset $\{m_i\}_{i \in I} \subseteq M$ and a set $\{f_i\}_{i \in I}$ of homomorphisms $f_i : M \to R_R$ such that $m = \sum_{i \in I} m_i f_i(m)$ for every $m \in M$, where $f_i(m) = 0$ for all but finitely many i.*
3. *There exist a generator system $\{m_i\}_{i \in I}$ of the module M and a set $\{f_i\}_{i \in I}$ of homomorphisms $f_i : M \to R_R$ such that $m = \sum_{i \in I} m_i f_i(m)$ for every $m \in M$, where $f_i(m) = 0$ for all but finitely many i.*

Proof $(1) \Rightarrow (2)$. In view of Theorem 1.57, we can assume that $M \oplus P = Q_R$, where Q_R is a free module with basis $\{x_i\}_{i \in I}$. Let $g_i : x_i R \to R_R$ be isomorphisms with $g(x_i) = 1$, $t : Q \to M$ be the projection with kernel P, $m_i \equiv t(x_i)$, $h_i : Q \to x_i R$ be natural projections, and let $f_i \equiv g_i h_i |_M : M \to R_R$. We consider $m \in M$. There exists a finite subset $J \subseteq I$ with $m = \sum_{i \in J} x_i a_i$. Since

$$\sum_{i \in J} m_i f_i(m) = \sum_{i \in J} m_i g_i(h_i(m)) = \sum_{i \in J} m_i g_i(x_i a_i) = \sum_{i \in J} m_i a_i = m,$$

the sets $\{m_i\}_{i \in I}$ and $\{f_i\}_{i \in I}$ satisfy the required properties.

$(2) \Rightarrow (3)$. Since $m = \sum_{i \in I} m_i f_i(m)$ for all $m \in M$, the set $\{m_i\}_{i \in I}$ generates M.

$(3) \Rightarrow (1)$. Let Q_R be a free module with basis $\{x_i\}_{i \in I}$, $u_i : R_R \to x_i R$ be isomorphisms with $u_i(1) = x_i$, and let $t : Q \to M$ be an epimorphism such that $t(x_i) = m_i$ for every $i \in I$. We define a homomorphism $f : M \to Q$ by the relation $f(m) = \sum_{i \in I} u_i(f_i(m)) = \sum_{i \in I} x_i f_i(m)$. We can verify that f is well defined. Then

$$(tf)(m) = t\left(\sum_{i \in I} x_i f_i(m)\right) = \sum_{i \in I} m_i f_i(m) = m.$$

Therefore, $tf = 1_M$ and module M is isomorphic to a direct summand of the free module Q. By Theorem 1.57, M is a projective module. $\square$

1.8 Injective Envelope and Quasi-Injective Modules

It may be shown using the Zorn lemma that every module M has a maximal essential extension E, which can also be shown to be a minimal injective extension of M. This minimal injective extension E of M is unique up to isomorphism and is called the *injective envelope* of M, denoted as $E(M)$. In what follows, we state a sequence of results that leads to the preceding statement.

Theorem 1.59 *Let M be an R-module. Then M is injective if and only if it has no proper essential extensions.*

Theorem 1.60 *Let M be an R-module, and let N be an R-module containing M. Then the following statements are equivalent:*

1. *N is a maximal essential extension of M.*
2. *N is a minimal injective extension of M.*

Theorem 1.61 *If $M = \oplus_{i \in I} M_i$ and for every $i \in I$, the modules M_i and $\oplus_{j \neq i} M_j$ are M_i-injective, then M is a quasi-injective module.*

Theorem 1.62 *A module M is a quasi-injective module if and only if M is invariant under any endomorphism of its injective envelope.*

Proof Suppose M is a module that is invariant under endomorphisms of its injective envelope. Suppose N is a submodule and we have a homomorphism $f : N \rightarrow M$. Clearly, we can extend it to a homomorphism $g : E(M) \rightarrow E(M)$. Since by assumption M is invariant under g, we have that $g|_M : M \rightarrow M$ extends f. This shows M is quasi-injective.

Conversely, suppose that M is quasi-injective. Consider an endomorphism f of $E(M)$. Let $L = \{m \in M : f(m) \in M\}$. Then L is a submodule of M. Since M is quasi-injective, we can extend the homomorphism $f|_L$ to an endomorphism g of M, and g can be extended to a homomorphism h of $E(M)$. We claim that $h|_M = f|_M$. Assume to the contrary that $(h - f)(M) \neq 0$. Then since M is essential in $E(M)$, we obtain that $(h - f)(M) \cap M \neq 0$. Then there exist $m, m' \in M$ such that $(h - f)(m) = m'$. Then we obtain that $h(m) - m' = f(m)$, and this yields that $m \in L$. Therefore, $m' = 0$, and this shows that M is invariant under endomorphisms of its injective envelope. $\square$

It is known that M is a quasi-injective right R-module if and only if M is a quasi-injective $R/r(M)$-module, where $r(M)$ is the annihilator of M. If M is a quasi-injective (e.g. injective) module, then M is an indecomposable module if and only if for every endomorphism f of M, at least one of the endomorphisms $f, 1 - f$ is an automorphism, i.e. the endomorphism ring End(M) of the module M is a local ring. In this case, all submodules of the module M are uniform.

Theorem 1.63 *For any module M, the following are equivalent.*

1. *Every idempotent endomorphism of each submodule in M can be extended to an endomorphism of M.*
2. *Every idempotent endomorphism of each submodule in M can be extended to idempotent endomorphism of M.*

3. M is an **idempotent-invariant** *module, i.e. $\alpha(M) \subseteq M$ for every idempotent endomorphism α of the injective envelope $E(M)$ of M.*
4. *Every submodule in M is an essential submodule of some direct summand of M, and for any two direct summands X, Y of M with $X \cap Y = 0$, we have that $X \oplus Y$ is a direct summand in M.*
5. *$M = \oplus_{i \in I}(M \cap E_i)$ for every the direct decomposition $E = \oplus_{i \in I} E_i$ of the injective envelope E of M.*
6. *For every submodule $X = X_1 \oplus X_2 \oplus \cdots \oplus X_n$ of M, there exists a direct decomposition $M = M_1 \oplus \cdots \oplus M_n \oplus Y$ of the module M such that M_i is an essential extension of X_i, $i = 1, 2, \ldots n$.*
7. *For every submodule $X = M_1 \oplus \cdots \oplus M_n$ of M such that $M_1, \ldots, M_n$ are closed submodules in M, we have $X = M$.*

Proof The equivalences (2) if and only if (3) if and only if (4) are proved in [70]; also see [113].

The equivalences (3) if and only if (5) and (4) if and only if (6) if and only if (7) are directly verified; also see [113].

The implication $(2) \Rightarrow (1)$ is obvious.

$(1) \Rightarrow (3)$. Let $X = \{m \in M \mid \alpha(m) \in M\}$, and let $f = \alpha|_X : X \to M$ be the restriction of the endomorphism α to the module X. Since $\alpha = \alpha^2$, we have $f(X) \subseteq X$. By (a), the homomorphism f can be extended to some endomorphism g of M. Since Q is an injective module, the endomorphism g of M can be extended to some endomorphism β of Q.

If $(\alpha - \beta)(M) = 0$, then $\alpha(M) = \beta(M) \subseteq M$, which is required.

Now we assume that $(\alpha - \beta)(M) \neq 0$. Since Q is an essential extension of M and $X = \{m \in M : \alpha(m) \in M\}$, we have that X is an essential submodule in Q. Then $X \cap (\alpha - \beta)(M)$ is a nonzero submodule in M, since Q is an essential extension of the module X. Let $0 \neq x = (\alpha - \beta)(m) \in X \cap (\alpha - \beta)(M)$, where $m \in M$. Since $\alpha(m) = (\alpha - \beta)(m) + \beta(m) = x + \beta(m) \in M$, we have $m \in X$. Therefore, $(\alpha - \beta)(m) = 0$ and $x = 0$. This is a contradiction. □

A module M is said to be **quasi-continuous** or π-injective if any of the preceding equivalent conditions hold for M.

Definition 1.64 A module M is said to satisfy the property C_1 (or it is called a **CS module**) if every submodule of M is an essential submodule of some direct summand of M.

Definition 1.65 A module M is said to satisfy the property C_2 if any submodule isomorphic to a direct summand of M is itself a direct summand of M.

Definition 1.66 A module M is said to satisfy the property C_3 if for any two direct summands X, Y of M with $X \cap Y = 0$, we have that $X \oplus Y$ is a direct summand of M.

Theorem 1.67 *Let M be a right R-module.*

1. *M is quasi-continuous if and only if M is a quasi-continuous $R/r(M)$-module, where $r(M)$ is the annihilator of M.*
2. *If M is a uniform module, then all submodules of M are quasi-continuous finite-dimensional modules.*
3. *If M is a quasi-continuous finite-dimensional module, then M is a finite direct sum of uniform modules.*
4. *If M is a quasi-continuous module, then the ring $\mathrm{End}\, M/\Delta(M)$ is isomorphic to the direct product of a right self-injective von Neumann regular ring and a reduced ring.*

Theorem 1.68 *Let R be any ring.*

1. *If R is right quasi-continuous domain, then R is a right Ore domain.*
2. *If R is a right nonsingular right quasi-continuous ring, then R is a Baer ring.*
3. *If R is a right nonsingular right quasi-continuous ring, then R is a direct product of a right self-injective von Neumann regular ring and a right quasi-continuous reduced Baer ring.*

Theorem 1.69 *For any module M, the following are equivalent.*

1. *Every submodule of M isomorphic to a closed submodule of M is a direct summand of M.*
2. *M is a quasi-continuous module, and every submodule of M, which is isomorphic to a direct summand of M, is a direct summand of M.*
3. *M is a quasi-continuous module, and for every endomorphism f of M such that $\mathrm{Ker}\, f$ is a closed submodule in M, the modules $\mathrm{Ker}\, f$ and $f(M)$ are direct summands of M.*
4. *M is a quasi-continuous module, and for every endomorphism f of M such that $\mathrm{Ker}\, f$ is a direct summand of M, the module $f(M)$ is a direct summand of M.*

Proof $(1) \Rightarrow (2)$. Let Q_1 and Q_2 be two closed submodules of M such that $Q_1 \cap Q_2 = 0$, and M is an essential extension of $Q_1 \oplus Q_2$. Since Q_1 is a closed submodule of M, there is a direct decomposition $M = Q_1 \oplus D$. Let $h \colon M \to D$ be the projection with kernel Q_1. Then the submodule $h(Q_2)$ of D is isomorphic to the closed submodule Q_2 of M. Therefore, $h(Q_2)$ is a direct summand in D. In addition, D is an essential extension of $D \cap (Q_1 \oplus Q_2) = h(D \cap (Q_1 \oplus Q_2)) = h(Q_2)$. Therefore, $D = h(Q_2)$ and $M = Q_1 \oplus D = Q_1 \oplus h(Q_2) = Q_1 \oplus Q_2$. By Theorem 1.63, M is a quasi-continuous module. Since $\operatorname{Ker} f$ is a closed submodule of M, there is a direct decomposition $M = \operatorname{Ker} f \oplus N$. Therefore, $f(M) \cong M/\operatorname{Ker} f \cong N$. Then the module $f(M)$ is isomorphic to a closed submodule of M. Therefore, $f(M)$ is a direct summand in M.

The implications $(3) \Rightarrow (2)$ and $(3) \Rightarrow (4)$ are directly verified.

$(2) \Rightarrow (1)$. The proof follows from the fact that every closed submodule of the quasi-continuous module M is a direct summand of M.

$(4) \Rightarrow (1)$. Let G be a closed submodule of M, let $g \colon G \to M$ be a monomorphism, and let $N \equiv g(G)$. By Theorem 1.63, there is a direct decomposition $M = G \oplus H$. Let $h \colon M \to G$ be the projection with kernel H, and let $f = gh \in \operatorname{End} M$. Then $\operatorname{Ker} f = H$ is a direct summand of $\varphi(M)$ and $N = f(M)$. By assumption, N is a direct summand of M. □

A module M is said to be a continuous module if any of the preceding equivalent conditions hold for M.

Theorem 1.70 *Every quasi-injective module is a continuous module.*

Proof Let M be a quasi-injective module. First, we prove that M is a continuous module. Let G be a closed submodule of M, $g \colon G \to M$ be a monomorphism, and let $N \equiv g(G)$. By Theorem 1.63, the quasi-injective module M is quasi-continuous. Therefore, there is a direct decomposition $M = G \oplus H$. Let $h \colon M \to G$ be a natural projection. Since M is a quasi-injective module, the homomorphism $g^{-1} \colon N \to G$ can be extended to an endomorphism f of M. We set $u = ghf \in \operatorname{Hom}(M, N)$. For $x \in N$, we obtain $u(x) = g(h(g^{-1}(x))) = g(g^{-1}(x)) = x$. Therefore, N is a direct summand of M. □

Definition 1.71 For a module M, if there exists an epimorphism $f \colon P \to M$ such that P is a projective module and $\operatorname{Ker} f$ is small in P, then this epimorphism is called a projective cover of M; in this case, the module P is generally called the projective cover of M.

Theorem 1.72 *Let M be a module with projective cover $f \colon P \to M$.*

1. *If Q is a projective module, and there exists an epimorphism $g \colon Q \to M$, then there exists an epimorphism $h \colon Q \to P$ such that $g = fh$, $Q = \operatorname{Ker} h \oplus \overline{P}$, and the restriction h to $\overline{P}$ is an isomorphism from the direct summand $\overline{P}$ of Q onto P; in particular, P is isomorphic to a direct summand of Q.*

2. *If $f^*\colon P^* \to M$ is one more projective cover of M, then there exist isomorphisms $h\colon P^* \to P$ and $h^* = h^{-1}\colon P \to P^*$ such that $f^* = fh$ and $f = f^*h^*$. In particular, any two projective covers of M are isomorphic to each other.*

Proof (1) Since Q is projective, there exists a homomorphism $h\colon Q \to P$ with $g = fh$. Since f is an epimorphism, $h(Q) + \mathrm{Ker}(f) = P$. Since $\mathrm{Ker}\, f$ is small in P, we have $h(Q) = P$. Therefore, h is an epimorphism onto the projective module P and there exists a direct decomposition $Q = \mathrm{Ker}\, h \oplus \overline{P}$.

(2) By (1), there exist epimorphisms $h\colon P^* \to P$ and $h^*\colon P \to P^*$ such that

$$f^* = fh, \quad f = f^*h^*, \quad P^* = \mathrm{Ker}\, h \oplus \overline{P},$$
$$P = \mathrm{Ker}(h^*) \oplus \overline{P^*}.$$

Therefore, $f^* = f^*h^*h$ and $(1 - h^*h)(P^*) \subseteq \mathrm{Ker}(f^*)$. Then the module $(1 - h^*h)(P^*)$ is small in P^*. In addition, $\mathrm{Ker}(h) = (1 - h^*h)(\mathrm{Ker}\, h \subseteq (1 - h^*h)(P^*)$. Therefore, $\mathrm{Ker}\, h$ is a small direct summand of P^*. Then $\mathrm{Ker}\, h = 0$, h is an isomorphism, $h^*h = 1$. Similarly, h^* is an isomorphism and $hh^* = 1$. □

Definition 1.73 A module M is said to be a hereditary module if all submodules of M are projective. The module M is said to be a semihereditary module if all finitely generated submodules of M are projective.

Definition 1.74 A ring R is called a right perfect ring if every right R-module has a projective cover.

Definition 1.75 A ring R is called a semiperfect ring if every finitely generated right (or left) R-module has a projective cover.

Theorem 1.76 *For a ring R, the following are equivalent:*

1. *R is left perfect.*
2. *R has DCC on principal right ideals.*
3. *$R/J(R)$ is semisimple Artinian, and $J(R)$ is left T-nilpotent (that is, for every sequence $a_1, a_2, \dots$ in $J(R)$, $a_1 a_2 \dots a_n = 0$ for some positive integer n).*
4. *$R/J(R)$ is semisimple Artinian, and every nonzero left R-module contains a maximal submodule.*
5. *R contains no infinite set of orthogonal idempotents and every nonzero right R-module contains a minimal submodule.*

1.9 Flat Modules

Definition 1.77 Let R be a ring, and let X_R, $_RY$ be right and left R-modules, respectively, $X \times Y$ be the cartesian product, let F be the free $\mathbb{Z}$-module with basis indexed by the set $X \times Y$, and let H be the subgroup of F generated by all elements of the form

$$(x + u, y) - (x, y) - (u, y), \quad (x, y + v) - (x, y) - (x, v), \quad (xa, y) - (x, ay),$$

$$x, u \in X, \quad y, v \in Y, \quad a \in R.$$

The abelian group F/H is called the **tensor product** of the modules X and Y; it is denoted by $X \otimes _RY$. We write $X \otimes Y$ instead of $X \otimes _RY$ if it is clear which the ring R is considered. The image of the pair (x, y) for the natural mapping $X \times Y \to X \otimes Y$ is denoted by $x \otimes y$.

If $_BX_R$ and $_RY_C$ are two bimodules, then the group $X \otimes_R Y$ can be naturally turned into a B-C-bimodule, in which multiplication by elements $b \in B$ and $c \in C$ is defined by the relations $b\left(\sum x_i y_i\right) = \sum b x_i y_i$ and $\left(\sum x_i y_i\right)c = \sum x_i y_i c$. In particular, $X \otimes_R Y$ is an $\text{End}(X)$-$\text{End}(Y)$-bimodule and $X \otimes_R R$ is a right R-module.

Lemma 1.78 *For any element x of the group $X \otimes _RY$, there exists a finite set of subscripts I with $x = \sum_{i \in I} x_i \otimes y_i$.*

1. If $x, u \in X$, $y, v \in Y$ and $a \in R$, then

$$(x + u) \otimes y = x \otimes y + u \otimes y, \quad x \otimes (y + v) = x \otimes y + x \otimes v,$$

$$xa \otimes y = x \otimes ay.$$

2. If $\sum_{i=1}^{n} x_i \otimes y_i \in X \otimes _RY$, then $\sum_{i=1}^{n} x_i \otimes y_i = 0$ if and only if there exist two finite sets $\{\overline{X}_k \in X\}_{k=1}^{m}$ and $\{a_{ik} \in R\}$ such that

$$1 \leq i \leq n, \quad 1 \leq k \leq m, \quad x_i = \sum_{k=1}^{m} \overline{x}_k a_{ik}, \quad \sum_{i=1}^{n} a_{ik} y_i = 0 \quad \forall i, k.$$

3. For any two module homomorphisms $f : X_R \to M_R$ and $g : {}_RY \to {}_RN$, the relation $(f \otimes g)\left(\sum x_i \otimes y_i\right) = \sum f(x_i) \otimes g(y_i)$ correctly defines a group homomorphism $f \otimes g : X \otimes _RY \to M \otimes _RN$.

4. The canonical group epimorphism $h : X \otimes_R R \to X$ is an isomorphism from the natural right R-module $X \otimes _RR$ onto the module X_R.

5. If $X_R = \oplus_{i \in I} X_i$, then there exists a natural group isomorphism from the group $(\oplus_{i \in I} X_i) \otimes _RY$ onto the group $\oplus_{i \in I} (X_i \otimes _RY)$.

6. *If P and Q are two submodules of $_RY$, then the intersection in $X \otimes_R (P + Q)$ of canonical images of the modules $X \otimes_R P$ and $X \otimes_R Q$ coincides with the canonical image of $X \otimes_R (P \cap Q)$.*

7. *If $_BX_R$, $_RY_C$ and $_CZ_D$ are bimodules, then there exists natural B-D-bimodule isomorphism $_B((X \otimes_R Y) \otimes_C Z)_D \to {}_B(X \otimes_R (Y \otimes_C Z))_D$.*

8. *If $_BX_R$, $_RY_C$ and $_DZ_C$ are bimodules and $_BX'_R$ is a sub-bimodule of $_BX_R$, then there exist natural B-D-bimodule isomorphisms*

$$\alpha: \mathrm{Hom}((X \otimes_R Y)_C, Z_C) \to \mathrm{Hom}(X_R, \mathrm{Hom}(Y_C, Z_C)),$$

$$\beta: \mathrm{Hom}((X' \otimes_R Y)_C, Z_C) \to \mathrm{Hom}(X'_R, \mathrm{Hom}(Y_C, Z_C))$$

and natural B-D-bimodule homomorphisms

$$f: \mathrm{Hom}((X \otimes_R Y)_C, Z_C) \to \mathrm{Hom}((X' \otimes_R Y)_C, Z_C),$$

$$g: \mathrm{Hom}(X_R, \mathrm{Hom}(Y_C, Z_C)) \to \mathrm{Hom}(X'_R, \mathrm{Hom}(Y_C, Z_C)),$$

with $g\alpha = \beta f$. Therefore, if f is an epimorphism, then g is an epimorphism.

Definition 1.79 Let R be a ring, let X be a right R-module, and let X' be a submodule of X such that for every left R-module Y, a natural group homomorphism $X' \otimes_R Y \to X \otimes_R Y$ is a monomorphism. Then X' is called a pure submodule of X. (In this case, we can consider $X' \otimes_R Y$ as a subgroup of $X \otimes_R Y$.)

Definition 1.80 A sequence of module homomorphisms $\cdots \longrightarrow M_{n-1} \xrightarrow{f_{n-1}} M_n \xrightarrow{f_n} M_{n+1} \xrightarrow{f_{n+1}} \cdots$ is said to be exact in the term M_n if $f(M_{n-1}) = \mathrm{Ker}(f_n)$ for all n. A sequence of homomorphisms is said to be exact if it is exact in every term of it.

Definition 1.81 For a ring R, a right R-module X is called a Hattori torsion-free module or an H-torsion-free module if for any $a \in R$, a natural group epimorphism $X \otimes Ra \to Xa$ is an isomorphism.

Theorem 1.82 *1. The union of any ascending chain of pure submodules is a pure submodule.*

2. *Every direct summand and the union of any ascending chain of direct summands is a pure submodule.*

3. *If $_SX_R$ and $_SX'_R$ are two S-R-bimodules and X'_R is a pure submodule of X_R, then for every R-C-bimodule $_RY_C$, $(X' \otimes_R Y)_C$ is a pure submodule of $(X \otimes_R Y)_C$.*

4. If $0 \longrightarrow X_1 \xrightarrow{f} X_2 \xrightarrow{g} X_3 \longrightarrow 0$ is an exact sequence of right
   R-modules, then $X_1 \otimes Y \xrightarrow{f \otimes 1} X_2 \otimes Y \xrightarrow{g \otimes 1} X_3 \otimes Y \longrightarrow 0$ is an exact
   sequence of abelian groups for any left R-module Y. In particular, $g \otimes 1$ is
   an epimorphism and $X_3 \otimes Y \cong (X_2 \otimes Y)/(f \otimes 1)(X_1 \otimes Y)$.

Theorem 1.83 *For a ring R and a right R-module X, the following conditions
are equivalent.*

1. *X is Hattori torsion-free module.*
2. *Every cyclic submodule in X is contained in some Hattori torsion-free
   submodule in X.*
3. *For any $x \in X$ and $a \in R$ with $xa = 0$, there exist $x_1, \ldots, x_n \in X$ and
   $a_1, \ldots, a_n \in R$ such that $m = \sum_{i=1}^{n} x_i a_i$ and $a_i a = 0$ for all i.*

Proof The equivalence (1) if and only if (2) is directly verified.

$(1) \Rightarrow (3)$. We assume that $xa = 0$, where $x \in X$ and $a \in R$. By assumption,
$x \otimes a = 0$. By Lemma 1.78(2), $x = \sum_{j=1}^{k} x_j a_j$, where $x_j \in X$, $a_j \in R$, and
$a_j a = 0$ for all j. Therefore, X is H-torsion-free.

$(3) \Rightarrow (1)$. We assume that $\sum_{i=1}^{n} x_i b_i a = 0$, where $x_i \in X$ and $b_i \in R$.
We set $x = \sum_{i=1}^{n} x_i b_i \in X$. Then $xa = 0$. Since X is H-torsion-free,
$x = \sum_{j=1}^{k} x_j a_j$, where $x_j \in X$, $a_j \in R$ and $a_j a = 0$ for all j. Then

$$\sum_{i=1}^{n} x_i \otimes b_i a = x \otimes a = \sum_{j=1}^{k} x_j a_j \otimes a = \sum_{j=1}^{k} x_j \otimes a_j a = \sum_{j=1}^{k} x_j \otimes 0 = 0. \quad \square$$

Let R be a ring, let X be a right R-module, and let X' be a submodule in
X. Then X/X' is an Hattori torsion-free module if and only if for any $x \in X$
and $a \in R$ with $xa \in X'$, there exist $x_1, \ldots, x_n \in X$ and $a_1, \ldots, a_n \in R$
such that $x = \sum_{i=1}^{n} x_i a_i \in X'$ and $a_i a = 0$ for all i. As a consequence, it
follows that for a module X, all submodules of X are Hattori torsion-free if
and only if all cyclic submodules in X are Hattori torsion-free modules.

For a ring R, a right R-module X is said to be flat if for any monomorphism
of left R-modules $Y' \rightarrow Y$, a natural group homomorphism $X \otimes_R Y' \rightarrow
X \otimes_R Y$ is a monomorphism. It is clear that every flat module is Hattori
torsion-free.

Theorem 1.84 *For a ring R and a right R-module X, the following conditions
are equivalent.*

1. *X is flat.*
2. *X is a direct summand of a direct sum of flat modules.*

3. *For every finitely generated left ideal Y in R, the canonical group epimorphism $X \otimes_R Y \to XY$ is an isomorphism.*
4. *For every left ideal Y in R, the canonical group epimorphism $X \otimes_R Y \to XY$ is an isomorphism.*
5. *For any $x_1, \ldots, x_n \in X$ and $y_1, \ldots, y_n \in R$ with $\sum_{i=1}^{n} x_i y_i = 0$, there exist $\overline{X}_1, \ldots, \overline{X}_m \in X$ and $a_{ik} \in R$ such that $1 \leq i \leq n$, $1 \leq k \leq m$, $x_i = \sum_{k=1}^{m} \overline{X}_k a_{ik}$, and $\sum_{i=1}^{n} a_{ik} y_i = 0$ for all i and k.*
6. *X is Hattori torsion-free and $XB \cap XC = X(B \cap C)$ for any two left ideals B and C in R.*
7. *X is Hattori torsion-free and $XB \cap XC = X(B \cap C)$ for any two finitely generated left ideals B and C in R.*
8. *Every finitely generated submodule of X is contained in some flat submodule in X.*

Proof The equivalence of conditions (1), (3) and (4) is verified easily.

The implications $(1) \Rightarrow (2)$, $(6) \Rightarrow (7)$ and $(1) \Rightarrow (8)$ are obvious.

The implication $(2) \Rightarrow (1)$ follows easily.

$(4) \Rightarrow (5)$. We set $Y = \sum_{i=1}^{n} R y_i \subseteq R$. Since $\sum_{i=1}^{n} x_i y_i = 0$, it follows from (e') that $\sum_{i=1}^{n} x_i \otimes y_i = 0$. Now we use Lemma 1.78(2).

$(5) \Rightarrow (6)$. Let Y be a left ideal in R, let $h \colon X \otimes Y \to XY$ be the canonical group epimorphism, and let $\sum_{i=1}^{n} x_i \otimes y_i \in \operatorname{Ker} h$. Then $\sum_{i=1}^{n} x_i y_i = 0$, and there exist $\overline{X}_1, \ldots, \overline{X}_m \in X$ and $a_{ik} \in R$ such that

$$1 \leq i \leq n, \quad 1 \leq k \leq m, \quad x_i = \sum_{k=1}^{m} \overline{X}_k a_{ik}, \quad \sum_{i=1}^{n} a_{ik} y_i = 0,$$

$$\sum_{i=1}^{n} x_i \otimes y_i = \sum_{i=1}^{n} \sum_{k=1}^{m} \overline{X}_i a_{ik} \otimes y_i = \sum_{k=1}^{m} \sum_{i=1}^{n} \overline{X}_i \otimes a_{ik} y_i = \sum_{i=1}^{n} \overline{X}_i \otimes 0 = 0.$$

Therefore, h is an isomorphism.

$(7) \Rightarrow (3)$. Let $k \in \mathbb{N}$, and let Y be a k-generated left ideal. We have to prove that the canonical group epimorphism $f_Y \colon X \otimes Y \to XY$ is an isomorphism. We use the induction on k. For $k = 1$, the assertion follows from the property that M is H-torsion-free. Let $Y = B + C$ be a k-generated left ideal, where B is a $(k-1)$-generated left ideal and C is a principal left ideal. Let

$$f_B \colon X \otimes B \to XB, \quad f_C \colon X \otimes C \to XC, \quad w \colon X \otimes A \to XR,$$
$$h_B \colon X \otimes B \to X \otimes Y, \quad h_C \colon X \otimes C \to X \otimes Y, \quad g \colon X \otimes Y \to X \otimes R$$

be canonical group homomorphisms, and let

$$p = \sum x_i \otimes (b_i - c_i) = \sum x_i \otimes b_i - \sum x_i \otimes c_i \in \operatorname{Ker} f_Y,$$

$b_i \in B$, $c_i \in C$, $q \equiv \sum x_i c_i \in XB$, $s \in X \otimes B$, $\sum x_i \otimes b_i = h_B(s)$.

Since $0 = f_Y(p) = \sum x_i(b_i - c_i)$, we have $q = \sum x_i c_i \in XB \cap XC$. By assumption, $XB \cap XC = X(B \cap C)$. Therefore, $q = \sum u_j d_j$ for some $u_j \in X$ and $d_j \in B \cap C$. Let $v \in X \otimes B$, and let $\sum u_j \otimes d_j = h_B(v)$. Then

$$0 = f_Y \left(\sum u_j \otimes d_j - \sum x_i \otimes b_i \right) = wg(h_B(v - s)) = f_B(v - s).$$

Therefore, $v = s$, since f_B is a monomorphism by the induction hypothesis. Then

$$\sum x_i \otimes b_i = h_B(s) = h_B(v) = \sum u_j \otimes d_j.$$

Similarly, we have $\sum x_i \otimes c_i = \sum u_j \otimes d_j$. Therefore, $p = \sum x_i \otimes b_i - \sum x_i \otimes c_i = 0$, and f is an isomorphism.

(1) $\Rightarrow$ (6). The intersection in $X \otimes_R (B + C)$ of canonical images of the modules $X \otimes_R B$ and $X \otimes_R C$ coincides with the canonical image of $X \otimes_R (B \cap C)$. Since (1) and (4) are equivalent, we obtain the required assertion.

(8) $\Rightarrow$ (7). X is H-torsion-free. Let B and C be two finitely generated left ideals, and let

$$x = \sum_{i=1}^{n} y_i b_i = \sum_{j=1}^{m} z_j c_j \in XB \cap XC, \quad y_i, z_j \in X, \quad b_i \in B, \quad c_j \in C.$$

We denote by Y the submodule in X generated by all elements y_i and z_j. Since Y is finitely generated, it follows from the assumption that Y is contained in some flat submodule M in X. Therefore, $x \in MB \cap MC$. Since M is a flat module and the implication (1) $\Rightarrow$ (6) has been proved, $MB \cap MC = M(B \cap C) \subseteq X(B \cap C)$. Therefore, $x \in X(B \cap C)$ and $XB \cap XC = X(B \cap C)$. □

Theorem 1.85 *All free or projective modules are flat.*

Proof Clearly, all free cyclic modules are flat. Since every free module is a direct sum of free cyclic modules, it follows that all free modules are flat. Since every projective module is isomorphic to a direct summand of a free module, it follows that all projective modules are flat. □

Theorem 1.86 *Let R be a ring, let X be a flat right R-module, and let Y be a submodule of X. The following conditions are equivalent.*

1. *X/Y is a flat module.*
2. *For any finitely generated left ideal B in R, a natural group epimorphism $XB/YB \to (X/Y)B$ is an isomorphism.*

3. *For any left ideal B in R, a natural group epimorphism*
   $XB/YB \to (X/Y)B$ *is an isomorphism.*
4. $Y \cap XB = YB$ *for any finitely generated left ideal B in R.*
5. $Y \cap XB = QB$ *for any left ideal B in R.*

Proof The sequence $Y \otimes B \longrightarrow X \otimes B \longrightarrow (X/Y) \otimes M \longrightarrow 0$ is exact. Since X is a flat module, it follows that the canonical epimorphism $X \otimes B \to XB$ is an isomorphism that maps from $Y \otimes B$ onto YB. Therefore, $(X/Y)B \cong XB/YB$. We know that X/Y is flat if and only if the canonical epimorphism $(X/Y) \otimes B \to (X/Y)B$ is an isomorphism for any (finitely generated) left ideal B. This implies the equivalence of conditions (1), (2) and (3).

The equivalence of conditions (1), (4) and (5) follows from the equivalence of conditions (1), (2), (3) and the property that $(X/Y)B = (XB + Y)/Y \cong XB/(Y \cap XB)$.

We prove that Y is flat provided that (5) holds. Since X is flat, it follows that X is H-torsion-free and $XB \cap XC = X(B \cap C)$. This property and (5) imply that

$$YB \cap YC = Y \cap XB \cap XC = Y \cap X(B \cap C) = Y(B \cap C).$$

It is sufficient to prove that Y is H-torsion-free. Let $ya = 0$, where $y \in Y$ and $a \in R$. Since X is H-torsion-free, there exist $x_1, \ldots, x_n \in X$ and $d_1, \ldots, d_n \in R$ such that $y = \sum_{i=1}^{n} x_i d_i$ and $d_i a = 0$ for all i. Let $D = \sum_{i=1}^{n} Rd_i$. Then $y \in Y \cap XD$. Therefore, $y \in YD$ by (5). Then there exist $y_1, \ldots, y_n \in Q$ with $y = \sum_{i=1}^{n} y_i d_i$. Since $d_i a = 0$ for all i, we have that Q is H-torsion-free. $\qquad\square$

Theorem 1.87 *Let R be a ring, let X be a free right R-module, and let Y be a submodule of X. The following conditions are equivalent.*

1. X/Y *is a flat module.*
2. *For any $y \in Y$, there exists a homomorphism $h: X \to Y$ with $h(y) = y$.*
3. *For any finite subset $\{y_1, \ldots, y_n\}$ in Y, there exists a homomorphism*
   $h: X \to Y$ *such that $h(y_i) = y_i$ for all $y_1, \ldots, y_n$.*

Proof (1) $\Rightarrow$ (2). Let $\{x_i\}_{i \in I}$ be a basis of the free module X_R, and let $\{x_1, \ldots, x_n\}$ be a subset of this basis with $y = \sum_{i=1}^{n} x_i a_i$, $a_1, \ldots, a_n \in R$. We denote by B the left ideal $\sum_{i=1}^{n} Ra_i$. Since $y \in Y \cap XB$ and $Y \cap XB = YB$, there exist $y_1, \ldots, y_k \in Y$ and $b_1, \ldots, b_k \in B$ with $y = \sum_{i=1}^{k} y_i b_i$. There exists a homomorphism $h: X \to X$ such that $h(x_i) = y_i$ for all $i = 1, \ldots, k$ and $h(x_i) = 0$ for all $i \in I \setminus \{1, \ldots, k\}$. Then $h(X) \subseteq Y$ and $h(y) = y$.

$(2) \Rightarrow (3)$. We use the induction on n. For $n = 1$, the assertion follows from (2). We assume that $n > 1$. Let $f : X \to Y$ be a homomorphism with $f(y_n) = y_n$. By the induction hypothesis, there exists a homomorphism $\varphi : X \to Y$ such that $\varphi(y_i - f(y_i)) = y_i - f(y_i)$ for all $i = 1, \ldots, n-1$. Therefore, $(1_X - \varphi)(1_X - f)(y_n) = (1_X - \varphi)(y_n - y_n) = 0$. For any $i < n$, we have $(1_X - \varphi)(1_X - f)(y_i) = (1_X - \varphi)(y_i - f(y_i)) = 0$. We set $h = 1_X - (1_X - \varphi)(1_X - f) \in \operatorname{End} X$. Then h has the required properties.

The implication $(3) \Rightarrow (2)$ is obvious.

$(2) \Rightarrow (1)$. Let B be a finitely generated right ideal in R, and let $y = \sum_{i=1}^{n} x_i b_i \in Y \cap XB$, where $x_i \in X$ and $b_i \in B$. By assumption, there exists a homomorphism $h : X \to Y$ with $y = h(y)$. Then $y = h(y) = \sum_{i=1}^{n} h(x_i) b_i \in YB$. Therefore, $Y \cap XB = YB$. Thus, X/Y is flat. $\square$

Theorem 1.88 *For a ring R and an element $a \in R$, the following conditions are equivalent.*

1. *$(R/aR)_R$ is a Hattori torsion-free module.*
2. *The principal right ideal aR is generated by an idempotent.*
3. *R/aR is a projective R-module.*

Proof The implications $(2) \Rightarrow (1)$ and $(2) \Rightarrow (3)$ are directly verified.

$(1) \Rightarrow (2)$. Since $1 \cdot a \in aR$ and R/aR is an H-torsion free right R-module, it follows that there exist $b, x_1, \ldots, x_n, a_1, \ldots, a_n \in R$ such that $1 = ab + \sum_{i=1}^{n} x_i a_i$ and $a_i a = 0$ for all i. Then $a = \left(ab + \sum_{i=1}^{n} x_i a_i \right) a = aba$. Therefore, ab is an idempotent and $abR = aR$. $\square$

Definition 1.89 A ring R is called a von Neumann regular ring if for each $a \in R$, there exists an element $b \in R$ such that $aba = a$.

Theorem 1.90 *For a ring R, the following statements are equivalent.*

1. *R is von Neumann regular.*
2. *Each principal right (left) ideal of R is generated by an idempotent.*
3. *Each finitely generated right (left) ideal of R is generated by an idempotent.*

Theorem 1.91 *For a ring R, the following conditions are equivalent.*

1. *R is von Neumann regular.*
2. *All right R-modules and all left R-modules are flat.*
3. *All cyclic right R-modules are flat.*
4. *For any element $a \in R$, the cyclic right R-module R/aR is Hattoti torsion free.*

Proof The implications $(2) \Rightarrow (3)$ and $(3) \Rightarrow (4)$ are obvious. The implication $(4) \Rightarrow (1)$ follows easily.

$(1) \Rightarrow (2)$. It is sufficient to prove that every right R-module X is flat. It is sufficient to prove that for any finitely generated left ideal Y, the canonical group epimorphism $f : X \otimes_R Y \to XY$ is an isomorphism. Let $p = \sum_{i=1}^{n} x_i \otimes y_i \in \mathrm{Ker}\, f$. Then $\sum_{i=1}^{n} x_i y_i = 0$. Since R is von Neumann regular, there exists an idempotent $e \in R$ with $Y = Re$. Then

$$p = \sum_{i=1}^{n} x_i \otimes y_i e = \left(\sum_{i=1}^{n} x_i y_i \right) \otimes e = 0 \otimes e = 0, \quad \mathrm{Ker}\, f = 0. \qquad \square$$

Theorem 1.92 *For a ring R, the following conditions are equivalent.*

1. *All submodules of flat right R-modules are flat.*
2. *All submodules of flat left R-modules are flat.*
3. *All right ideals of R are flat.*
4. *All left ideals of R are flat.*
5. *All finitely generated right ideals of R are flat.*
6. *All finitely generated left ideals of R are flat.*
7. *For any finitely generated right ideal X of R and each finitely generated left ideal Y of R, a natural group homomorphism $X \otimes Y \to XY$ is an isomorphism.*

Proof Since condition (7) is left-right symmetrical, it is sufficient to prove the equivalence of conditions (1), (3), (5) and (7).

The implication $(1) \Rightarrow (3)$ follows from the property that R_R is a flat right module.

The equivalence (3) if and only if (5) and (5) if and only if (7) follows easily.

$(7) \Rightarrow (1)$. Let M be a flat right R-module, let X be a submodule in M, let $f : X \to M$ be a natural embedding, and let Y be a finitely generated left ideal in R. We have that $_R Y$ is flat. Therefore, $f \otimes 1$ is a monomorphism. We assume that $\sum_{i=1}^{n} x_i y_i = 0$, where $x_1, \ldots, x_n \in X$ and $y_1, \ldots, y_n \in Y$. Since M_R is flat, it follows that a natural group epimorphism $M \otimes Y \to MY$ is an isomorphism. Therefore, $(f \otimes 1) \left(\sum_{i=1}^{n} x_i \otimes y_i \right) = 0$. Since $f \otimes 1$ is a monomorphism, $\sum_{i=1}^{n} x_i y_i = 0$. Therefore, a natural group epimorphism $X \otimes Y \to XY$ is an isomorphism. Thus, X is flat. $\qquad \square$

Theorem 1.93 *Let Q be a ring, and let R be a unitary subring in Q.*

1. *If M is a finitely generated flat right R-module and $M \otimes_R Q$ is a projective right Q-module, then M is a projective R-module.*

2. *If Q is a semisimple ring, then every finitely generated flat right or left R-module is projective.*
3. *If Q is a semisimple ring and all finitely generated right ideals of R are flat, then R is right and left semihereditary.*

Proof (**1**) There exists an exact sequence

$$0 \to X \to Y \to M \longrightarrow 0,$$

where Y is a finitely generated free module. Since M is a flat module, the sequence of abelian groups

$$0 \to X \otimes_R Q \to Y \otimes_R Q \to M \otimes_R Q \longrightarrow 0$$

is exact. Since $M \otimes Q$ is a projective Q-module, the module $(X \otimes Q)_Q$ is isomorphic to a direct summand in $(Y \otimes Q)Q$. We have that X_R is flat, since Y and M are flat R-modules. Let $\{q_1 \otimes 1, \ldots, q_n \otimes 1\}$ be a generator system for $(X \otimes Q)_Q$. So there exists a homomorphism $f : Y \to X$ such that $f(q_i) = q_i$ for any i. We consider any $q \in X$. Then $q \otimes 1 = \sum_{i=1}^{n}(q_i \otimes 1)b_i$, where $b_i \in Q$. Since X_R is a flat module, $X \cong X \otimes R \to X \otimes Q$ is a monomorphism. In the group $(X \otimes Q)_Q$, we have

$$f(q) \otimes 1 = (f \otimes 1)(q \otimes 1) = \sum (f \otimes 1)(q_i \otimes b_i) = \sum f(q_i) \otimes b_i$$

$$= \sum q_i \otimes b_i = q \otimes 1.$$

Then $f(q) = q$ for all $q \in X$; Therefore, X is a direct summand in Y. Then M is projective, since M is isomorphic to a direct summand of a projective module.

(**2**) The assertion follows from projectivity of all modules over a semisimple ring.

(**3**) The assertion follows from the property that all finitely generated left ideals in R are flat. $\qquad\square$

Theorem 1.94 *Every finitely presented flat module M is projective.*

Proof Let $M \cong F/Q$, where F is a finitely generated free module and Q is generated by $q_1, \ldots, q_n$. There exists a homomorphism $h \in \mathrm{Hom}(F, Q)$ such that $h(q_i) = q_i$ for all i. Then $h(q) = q$ for any $q \in Q$. Therefore, M is isomorphic to a direct summand of the free module F. $\qquad\square$

Theorem 1.95 *Every finitely presented flat module M with projective cover P is projective and $M = P$.*

Proof Let Q be a small submodule in P, let $P/Q \cong M$, and let N be a cyclic submodule in Q. It is sufficient to prove that $N = 0$. There exists a homomorphism $f: P \to Q$ with $(1 - f)(N) = 0$. Since $f(P)$ is small in P and $P = f(P) + (1 - f)(P)$, we have $(1 - f)(P) = P$. Since $1 - f$ is an epimorphism onto the projective module P, we have that $\mathrm{Ker}(1 - f)$ is a direct summand of P. If $x \in \mathrm{Ker}(1 - f)$, then $x = f(x) \in Q$. Then $\mathrm{Ker}(1 - f) \subseteq Q$, and $\mathrm{Ker}(1 - f)$ is small in P. Since $\mathrm{Ker}(1 - f)$ is a direct summand in P, we have $\mathrm{Ker}(1 - f) = 0$. Therefore, $N \subseteq \mathrm{Ker}(1 - f) = 0$. $\qquad\square$

1.10 Exchange Property of Modules

The notion of exchange property for modules was introduced by Crawley and Jónnson [21].

Definition 1.96 Let τ be a cardinal number. A module M is called a module with the τ-exchange property if for every module X and each direct decomposition $X = M' \oplus Y = \oplus_{i \in I} N_i$ such that $M' \cong M$ and $\mathrm{card}(I) \leq \tau$, there are submodules $N_i' \subseteq N_i$ $(i \in I)$ with $X = M' \oplus (\oplus_{i \in I} N_i')$.

It follows from the modular law that N_i' must be a direct summand of N_i for all i in the preceding definition.

Definition 1.97 A module M is said to satisfy the exchange property (or M is called an exchange module) if M is a module with the τ-exchange property for every cardinal number τ. A module M is called a module with the finite exchange property if M has the n-exchange property for every positive integer n.

We note that the direct sums in the definition of the exchange property are internal direct sums of submodules in X.

Theorem 1.98 *Every quasi-injective module satisfies the exchange property.*

Proof By Theorem 1.70, M is a continuous module. Let $X = M \oplus Y = \oplus_{i \in I} N_i$. We set $N_i' = N_i \cap Y$ and $N' = \oplus_{i \in I} N_i' \subseteq X$. By the Zorn lemma, there is a submodule P of X that is maximal with respect to the following properties:

1. $P = \oplus_{i \in I} P_i$, where $N_i' \subseteq P_i \subseteq N_i$ for all $i \in I$,
2. $M \cap P = 0$.

Let $h: X \to X/P$ be the natural epimorphism, let $j \in I$, and let $h(Q_j)$ be an arbitrary nonzero submodule of $h(N_j)$, where $P_j \subset Q_j \subseteq N_j$. Then P is a proper submodule of $Q_j + P = Q_j \oplus (\oplus_{i \neq j} P_i)$. It follows from the maximality of P that $M \cap (Q_j + P) \neq 0$. Since $M \cap P = 0$, we have $M \cap (Q_j + P) \not\subseteq P$. Then

$$0 \neq h(M \cap (Q_j + P)) = h(M) \cap h(Q_j) = (h(M) \cap h(N_j)) \cap h(Q_j).$$

Therefore, $h(M) \cap h(N_j)$ is an essential submodule of $h(N_j)$ for an arbitrary $j \in I$. Then $\oplus_{j \in I}(h(M) \cap h(N_j))$ is an essential submodule of $\oplus_{j \in I} h(N_j) = h(X)$. Therefore, $h(M)$ is an essential submodule of $h(X)$.

Let $f : M \oplus Y \to M$ be the projection with kernel Y. Since $N_i \cap \operatorname{Ker} f = N_i'$, the module N_i/N_i' is isomorphic to a submodule of M. In addition, M is an M-injective module. By Theorem 1.49, M is an N_i/N_i'-injective module. Since $X/N' \cong \oplus_{i \in I}(N_i/N_i')$, the module M is X/N'-injective by Theorem 1.49. Since $h(X)$ is a homomorphic image of X/N', the module M is $h(X)$-injective by Theorem 1.49. In addition, $M \cong h(M)$. Therefore, $h(M)$ is an $h(X)$-injective submodule of $h(X)$. By Theorem 1.53, $h(M)$ is a direct summand of $h(X)$. In addition, it was proved that $h(M)$ is an essential submodule of $h(X)$. Therefore, $h(M) = h(X)$. Then $X = M \oplus P$. □

Let M be a module, and let τ be a cardinal number.

Theorem 1.99 *If M is a module with the τ-exchange property, then every direct summand of M is a module with the τ-exchange property.*

Proof Let $M = M_1 \oplus M_2$, and let $X = M_1 \oplus Y = \oplus_{i \in I} N_i$, card $I \leq \tau$. Then $X \oplus M_2 = M \oplus Y = M_2 \oplus (\oplus_{i \in I} N_i)$. Since M is a module with the τ-exchange property, there are submodules $M_2' \subseteq M_2$ and $N_i' \subseteq N_i$, $i \in I$, such that $X \oplus M_2 = M \oplus M_2' \oplus (\oplus_{i \in I} N_i')$. Since $M_2' \subseteq M$ and $M_2' \cap M = 0$, we have $M_2' = 0$. Therefore, $X \oplus M_2 = M_1 \oplus M_2 \oplus (\oplus_{i \in I} N_i')$. It follows from $M_1 \oplus (\oplus_{i \in I} N_i') \subseteq X$ and the modular law that

$$X = X \cap (M_1 \oplus M_2 \oplus (\oplus_{i \in I} N_i')) = M_1 \oplus (\oplus_{i \in I} N_i') + X \cap M_2.$$

Since $X \cap M_2 = 0$, we have $X = M_1 \oplus (\oplus_{i \in I} N_i')$. Therefore, M_1 is a module with the τ-exchange property. □

Theorem 1.100 *If $M = M_1 \oplus \ldots \oplus M_k$ and all the modules M_i are modules with the τ-exchange property, then M is a module with the τ-exchange property.*

Proof We can assume that $k = 2$. Let $X = M_1 \oplus M_2 \oplus Y = \oplus_{i \in I} N_i$, where I is a set of cardinality τ. Since M_1 is a module with the τ-exchange property, $X = M_1 \oplus M_2 \oplus Y = M_1 \oplus (\oplus_{i \in I} S_i)$ with $S_i \subseteq N_i$, $i \in I$. Let $h : X \to \oplus_{i \in I} S_i$ be the projection with kernel M_1. Then $h(M_2) \oplus h(Y) = \oplus_{i \in I} S_i$ and

$X = M_1 \oplus h(M_2) \oplus h(Y)$. Since $M_2 \cap Kerh = 0$, the module $h(M_2)$ is isomorphic to M_2, whence $h(M_2)$ is a module with the τ-exchange property. Therefore, $h(M_2) \oplus h(Y) = h(M_2) \oplus (\oplus_{i \in I} T_i)$, where $T_i \subseteq S_i \subseteq N_i$ for all i. In addition, $M_1 \oplus h(M_2) = M_1 \oplus M_2$. Therefore,

$$X = M_1 \oplus h(M_2) \oplus h(Y) = M_1 \oplus h(M_2) \oplus (\oplus_{i \in I} T_i)$$
$$= M_1 \oplus M_2 \oplus (\oplus_{i \in I} T_i) = M \oplus (\oplus_{i \in I} T_i). \quad \square$$

Theorem 1.101 *Every direct summand of an exchange module is an exchange module, and each finite direct sum of exchange modules is an exchange module.*

Theorem 1.102 *M is a finitely generated exchange module if and only if M is a direct summand of a finite direct sum of finitely generated modules with the finite exchange property.*

Proof It is sufficient to consider the case where M is a direct summand of a finite direct sum of finitely generated modules with the finite exchange property. It is clear that the module M is finitely generated. By Theorem 1.101, M is a module with the finite exchange property. Let τ be a cardinal number, let X be a module, and let $X = M' \oplus Y = \oplus_{i \in I} N_i$ be a direct decomposition such that $M' \cong M$ and $card(I) \leq \tau$. Since M' is a finitely generated submodule of X, there is a finite subset J of A with $M' \subseteq \oplus_{j \in J} N_j$. Since M' is a module with the finite exchange property, there are submodules $N'_j \subseteq N_j$, $j \in J$, such that $\oplus_{j \in J} N_j = M' \oplus (\oplus_{j \in J} N'_j)$. For $i \in I \setminus J$, we set $N'_i \equiv N_i$. Then $X = M' \oplus (\oplus_{i \in I} N'_i)$. $\quad \square$

Theorem 1.103 *Let M be a module, let I be a set of cardinality τ, and let $M = B \oplus C \oplus E = (\oplus_{i \in I} D_i) \oplus E$, where B is a module with the τ-exchange property. Then there are submodules F_i of the modules D_i such that $M = B \oplus (\oplus_{i \in I} F_i) \oplus E$.*

Proof Let $h: M \rightarrow \oplus_{i \in I} D_i$ be the projection with kernel E. Since $M = B \oplus C \oplus E$, we have $\oplus_{i \in I} D_i = h(M) = h(B) \oplus h(C)$. In addition, $h(B)$ is a module with the τ-exchange property, since $h(B) \cong B$. Therefore, there are submodules F_i of the modules D_i $(i \in I)$ such that $h(M) = h(B) \oplus (\oplus_{i \in I} F_i)$, where $F_i \subseteq D_i$ $(i \in I)$. Therefore, $M = h(M) \oplus E = h(B) \oplus (\oplus_{i \in I} F_i) \oplus E$. Since $B \oplus E = h(B) \oplus E$, we have $M = B \oplus (\oplus_{i \in I} F_i) \oplus E$. $\quad \square$

Definition 1.104 A set $\{f_i\}_{i \in I}$ of module homomorphisms $X \rightarrow M$ is called a summable set if for every element x of X, we have $f_i(x) = 0$ for all but finitely many subscripts i. In this case, $\sum_{i \in I} f_i$ is a well-defined homomorphism from X into M.

Theorem 1.105 *Let M be a module, let $S = \text{End}(M)$, and let τ be a cardinal number. Then the following conditions are equivalent.*

1. *M is a module with the τ-exchange property.*
2. *For every direct decomposition $M \oplus B = \oplus_{i \in I} X_i$ such that* card $I \leq \tau$
   and $X_i \cong M$ for all $i \in I$, there are submodules $C_i \subseteq X_i$ with
   $M \oplus (\oplus_{i \in I} C_i) = \oplus_{i \in I} X_i.$
3. *For every summable set $\{f_i\}_{i \in I}$ of endomorphisms of M such that*
   card$(I) \leq \tau$ and $\sum_{i \in I} f_i = 1_M$, there is a summable set $\{e_i\}_{i \in I}$ of
   orthogonal idempotents of S such that $\sum_{i \in I} e_i = 1_M$ and $e_i \in Sf_i$ for all
   $i \in I$.

Proof The implication $(1) \Rightarrow (2)$ follows from the definition of a module with the τ-exchange property.

$(2) \Rightarrow (3)$. Let $\{X_i\}_{i \in I}$ be the set of isomorphic copies of M, let $X = \oplus_{i \in I} X_i$, let $\overline{M}$ be the submodule of X such that $\overline{M} = \{(f_i(m))_{i \in I} \mid m \in M\}$, let $f : M \to \overline{M}$ be the homomorphism with $f(m) = (f_i(m))_{i \in I}$, and let $g : X \to M$ be the homomorphism with $g((m_i)_{i \in I}) = \sum_{i \in I} m_i$. Then f is an isomorphism, $\overline{M} \cong M$ and $gf = 1_M$. Therefore, $\overline{M}$ is a direct summand of X. By (2), there are direct sum decompositions $X_i = B_i \oplus C_i$ such that

$$X = \overline{M} \oplus (\oplus_{i \in I} C_i) = (\oplus_{i \in I} B_i) \oplus (\oplus_{i \in I} C_i).$$

We set $U = \oplus_{i \in I} C_i$, $V = \oplus_{i \in I} B_i$. Then $X = U \oplus V = U \oplus \overline{M}$. Let $p : X \to V$ be the projection with kernel U. Since $\overline{M} \subseteq U + p(W)$ and $X = U + \overline{M}$, we have $X = U + p(\overline{M}) = U \oplus p(\overline{M}) = U \oplus V$. In addition, $p(\overline{M}) \subseteq V$. Therefore, $p(\overline{M}) = V$. Thus, the restriction of p to $\overline{M}$ is an isomorphism from $\overline{M}$ onto V. In other words, the projection $p : X \to \oplus_{i \in I} B_i$ with kernel $\oplus_{i \in I} C_i$ induces the isomorphism $t : \overline{M} \to \oplus_{i \in I} B_i$. Let $p_i : \oplus_{i \in I} B_i \to B_i$ be natural projections, let $t_i \equiv p_i t : \overline{M} \to B_i$, and let $e_i \equiv gt^{-1} t_i f \in S$. Then $e_i e_j = gt^{-1} t_i f gt^{-1} t_j f = gt^{-1} p_i t_j f$. Therefore, $(e_i)^2 = e_i$ and $e_i e_j = 0$ for $i \neq j$. Let $h : X_i \to B_i$ be the projection with kernel C_i. Then $t_i f = h_i f_i$, whence $e_i \in Sf_i$. Therefore, $\{e_i\}_{i \in I}$ is a summable set and $\sum_{i \in I} e_i = 1_M$.

$(3) \Rightarrow (1)$. Assume that $X = M \oplus B = \oplus_{i \in I} X_i$, $u_i : X \to X_i$ and $v : X \to M$ are corresponding projections, $f_i \equiv vu_i|_M \in S$, and card$(I) \leq \tau$. Then $\{f_i\}_{i \in I}$ is a summable set and $\sum_{i \in I} f_i = 1_M$. By assumption, there is a summable set $\{e_i\}_{i \in I}$ of idempotents of S such that $\sum_{i \in I} e_i = 1_M$ and $e_i = s_i f_i \in Sf_i$. We set $g_i \equiv e_i s_i vu_i : X \to M$. Then $X = M \oplus (\oplus_{i \in I} X_i \cap \text{Ker}(g_i))$. Therefore, $\{g_i\}_{i \in I}$ is a summable set of homomorphisms from X into M. We set $g \equiv \sum_{i \in I} g_i : X \to M$. Then $g_i|_M = e_i$, $g_i g_j = \delta_{ij}$, and $g|_M = 1_M$. It can be directly verified that $g^2 = g$ and $\text{Ker } g = \oplus_{i \in I} (X_i \cap \text{Ker}(g_i))$. $\square$

Theorem 1.106 *For a module M, the following conditions are equivalent.*

1. *M is an exchange module.*
2. *For every direct decomposition $M \oplus G = \oplus_{i \in I} F_i$ such that $F_i \cong M$ for all $i \in I$, there are submodules $H_i \subseteq F_i$ such that $M \oplus (\oplus_{i \in I} H_i) = \oplus_{i \in I} F_i$.*
3. *M is a module with the finite exchange property, and for every direct decomposition $M \oplus G = F = \oplus_{i \in I} F_i$ with $F_i \cong M$ for all $i \in I$, there is a submodule H of F such that H is a maximal element of the set of all submodules of F that have the following properties: $H = \oplus_{i \in I} H_i$, $H_i \subseteq F_i$, $H \cap M = 0$, and the natural monomorphism $M \to F/H$ is split.*

Proof The equivalence of (1) and (2) follows easily. The implication (1) + (2) ⇒ (3) is directly verified.

(3) ⇒ (2). Let $M \oplus G = F = \oplus_{i \in I} F_i$ be a direct decomposition such that $F_i \cong M$ for all $i \in I$. By (1), there is a submodule H of F such that H is a maximal element of the set $\mathcal{E}$ of all submodules of F such that $H = \oplus_{i \in I} H_i$, $H_i \subseteq F_i$, $H \cap M = 0$, and the natural monomorphism $\alpha \colon M \to F/H$ is split. Let $X = F/H$, $X_i = F_i/H_i$, and let $M^* = \alpha(M)$. There is a submodule Y of X such that $M^* \oplus Y = \oplus_{i \in I} X_i = X$. Since H is a maximal element of the set $\mathcal{E}$, the following assertion (*) holds for every $i \in I$: if Z_i is an arbitrary nonzero submodule of X_i such that $M^* \cap Z_i = 0$, then the sum $M^* \oplus Z_i$ is not a direct summand of X.

Let $J = \{1, \ldots, n\}$ be a finite subset of the set I. Since M^* is a module with the finite exchange property, we have that for every $k \in J$, there is a direct decomposition $X = M^* \oplus Z_k \oplus Z$ such that $Z_k \subseteq X_k$ and $Z \subseteq \oplus_{i \neq k} X_i$. It follows from assertion (*) that $Z_k = 0$. Then $X = M^* \oplus Z$. Therefore, the module X_k is isomorphic to a direct summand of M^* with the finite exchange property. Then X_k is a module with the finite exchange property. By Theorem 1.100, $\oplus_{k=1}^{n} X_k$ is a module with the finite exchange property. By Corollary 1.109, $S = \text{End}(\oplus_{k=1}^{n} X_k)$ is an exchange ring. Let $p \colon X \to Y$ be the projection with kernel M^*, let $e_k \colon X \to X_k$ be the projection with kernel $\oplus_{i \neq k} X_i$, and let $e \equiv e_1 + \cdots + e_n$. We identify the ring S with the ring $e(\text{End } X)e$. We choose an arbitrary integer $k \in J$ and $s \in S$. We set $a = sepe_k \in S$. Since S is an exchange ring and e is the identity element of S, there is an idempotent $f \in Sa$ with $e - f \in S(e-a)$. Then $f = ra = rsepe_k$, where $r = fr \in S$. We set $\varphi = e_k rsep$. Since $r = fr$, we have

$$\varphi^2 = e_k(rsepe_k)rsep = e_k frsep = e_k rsep = \varphi.$$

Therefore, $X = Ker\varphi \oplus \varphi(X)$. Since $M^* \subseteq Ker\varphi$, we have $X = M^* \oplus Z \oplus \varphi(X)$. It was proved that $X = M^* \oplus Z$. Therefore, $\varphi(X) = 0$. Then

$$f = f^2 = (ra^2) = rsep(e_k rsep)e_k = rsep\varphi e_k = 0.$$

Therefore, $e = e - f \in S(e-a) = S(e-sepe_k)$. Therefore, $e-sepe_k$ is a left invertible element of S. In addition, s is an arbitrary element of S, and e is the identity element of S. Then epe_k belongs to the Jacobson radical $J(S)$ of S. Therefore, $epe = \sum_{k=1}^{n} epe_k \in J(S)$. Since epe acts identically on the module $Y \cap (\oplus_{i \in J} X_i)$ and $epe \in J(S)$, we have $Y \cap (\oplus_{i \in J} X_i) = 0$ for an arbitrary finite subset J of the set I. Therefore, $Y = 0$. Then $M^* = M^* \oplus Y = X$, whence $F = M \oplus (\oplus_{i \in I} H_i)$. $\qquad\qquad\qquad\square$

Theorem 1.107 *For a right module M with endomorphism ring $R = \mathrm{End}(M)$, the following conditions are equivalent.*

1. *M is a module with the finite exchange property.*
2. *M is a direct summand of a finite direct sum of modules with the 2-exchange property.*
3. *For any two elements α and β of the ring R with $\alpha + \beta = 1$, there are two idempotents $e \in \alpha R$ and $f \in \beta R$ with $e + f = 1$.*
4. *For any two elements α and β of the ring R with $\alpha + \beta = 1$, there are two idempotents $e' \in R\alpha$ and $f' \in R\beta$ with $e' + f' = 1$.*

Proof The implication $(1) \Rightarrow (2)$ is obvious.

$(2) \Rightarrow (3)$. Clearly M is a module with the 2-exchange property. Let $X = M \oplus M$, $N_1 = \{(x,0) : x \in M\}$, $N_2 = \{(0,x) : x \in M\}$, $D = \{(x,x) : x \in M\}$, and let $M' = \{(\alpha(x), -\beta(x)) : x \in M\}$. Then $X = M' \oplus D = N_1 \oplus N_2$. Since $\mathrm{Ker}\,\alpha \cap \mathrm{Ker}\,\beta = 0$, the module M' is isomorphic to M. Since M is a module with the 2-exchange property, there are two submodules $N_i' \subseteq N_i$, $i = 1, 2$, with $X = M' \oplus N_1' \oplus N_2'$. Let x be an element of M. Then there is a unique decomposition (*): $(x,x) = (\alpha(y), -\beta(y)) + (x_1, 0) + (0, x_2)$, where $(x_1, 0) \in N_1'$ and $(0, x_2) \in N_2'$. By equating the first components in (*) and then the second components in (*), we obtain $x = \alpha(y) + x_1$ and $x = -\beta(y) + x_2$. Then $0 = \alpha(y) + x_1 + \beta(y) - x_2$, whence $x_2 - x_1 = (\alpha + \beta)(y) = y$. We define two endomorphisms α' and β' of M by $\alpha'(x) = x_2$ and $\beta'(x) = x_1$. It follows from the decompositions $(\alpha\alpha'(x), \alpha\alpha'(x)) = (\alpha\alpha'(x), \beta\alpha'(x)) + (0,0) + (0, \alpha'(x))$ and $(\beta\beta'(x), \beta\beta'(x)) = (-\alpha\beta'(x), \beta\beta'(x)) + (\beta'(x), 0) + (0,0)$, that $\alpha\alpha'\alpha = \alpha$ and $\beta\beta'\beta = \beta$. Therefore, $e \equiv \alpha\alpha'$ and $f \equiv \beta\beta'$ are idempotents of R, $e \in \alpha R$ and $f \in \beta R$. In addition, $\alpha'(x) - \beta'(x) = x_2 - x_1 = y$. Then

$$(e+f)(x) = \alpha\alpha'(x) + \beta\beta'(x) = \alpha(\beta'(x) + y) + \beta\beta'(x)$$
$$= (\alpha + \beta)\beta'(x) + \alpha(y)$$
$$= \beta'(x) + \alpha(y) = x_1 + (x - x_1) = x.$$

Therefore, $e + f = 1_M$ and (3) holds.

(3) $\Rightarrow$ (4). By assumption, there are two idempotents $e \in aR$ and $f \in bR$ with $e + f = 1$. Assume that $e = ar$ and $f = bs$, where $r, s \in R$. Replacing r with re and s with sf, we can assume $rar = r$, $rbs = 0$, $sbs = s$ and $sar = 0$. Let $r' \equiv 1 - sb + rb$, let $s' \equiv 1 - ra + sa$, let $e' \equiv r'a \in Ra$, and let $f' \equiv s'b \in Rb$. Then $r's = (1 - sb + rb)s = (s - sbs) + rbs = 0$ and $s'r = (1 - ra + sa)r = (r - rar) + sar = 0$. Since $ar + bs = e + f = 1$, we obtain $ar' = a(1 - sb) + arb = a(1 - sb) + (1 - bs)b = a(1 - sb) + b(1 - sb) = (a + b)(1 - sb) = 1 - sb$. Therefore, $r'ar' = r'$. Similarly, we have $bs' = 1 - ra$ and $s'bs' = s'$. Therefore, e' and f' are idempotents. Since $ab = ba$, we have $e' + f' = r'a + s'b = (a - sba + rba) + (b - rab + sab) = a + b = 1$.

(4) $\Rightarrow$ (3). The proof is similar to the proof of the implication (3) $\Rightarrow$ (4).

(3) $\Rightarrow$ (2). Let $X = M \oplus Y = N_1 \oplus N_2$ be a right module decomposition, $E = \operatorname{End} X$, $h: X \to M$ be the projection with kernel Y, t_1 and t_2 be two idempotents of E such that $t_1 + t_2 = 1_X$, $N_1 = t_1(X) = \operatorname{Ker} t_2$, and $N_2 = t_2(X) = \operatorname{Ker} t_1$. Since $\operatorname{End} M \cong hEh$ and $h = ht_1h + ht_2h$ is the identity element of hEh, it follows from (3) that there are orthogonal idempotents $v_i \in (hEh)ht_ih = hEht_ih$ $(i = 1, 2)$ with $v_1 + v_2 = h$. Then $v_i = \alpha_i t_i h$, where $\alpha_i = v_i \alpha_i \in hEh$. We set $w_i = t_i \alpha_i t_i$ $(i = 1, 2)$. Then w_i are orthogonal idempotents and $w_i(X) \subseteq N_i$ $(i = 1, 2)$. Therefore, $N_i = w_i(X) \oplus N_i'$, where $N_i' = N_i \cap \operatorname{Ker} w_i$. Then $X = w_1(X) \oplus w_2(X) \oplus N_1' \oplus N_2'$.

We prove that $X = M \oplus (N_1' \oplus N_2')$. Since $w_i h = t_i v_i$, we have $\alpha_i w_i h = \alpha_i t_i v_i = v_i^2 = v_i$. Let $y \in M \cap (N_1' \oplus N_2')$. Then $w_1(y) = 0 = w_2(y)$, whence we have

$$y = h(y) = v_1(y) + v_2(y) = \alpha_1 w_1 h(y) + \alpha_2 w_2 h(y)$$

$$= \alpha_1 w_1(y) + \alpha_2 w_2(y) = 0.$$

Therefore, $M \cap (N_1' \oplus N_2') = 0$. It remains to prove that any element x of X is contained in $M \oplus (N_1' \oplus N_2')$. Let $x = x_1 + x_2 + z$, where $x_i \in w_i(X)$ and $z \in N_1' \oplus N_2'$. We set $\delta_{ij} = 0$ for $i \neq j$ and $\delta_{ij} = 1$ for $i = j$. Then

$$w_j \alpha_i w_i = w_j(h\alpha_i)(t_i w_i) = (t_j v_j)(v_i \alpha_i)(t_i w_i) = \delta_{ij} w_i.$$

Therefore, $x_i - \alpha_i(x_i) \in N_i'$ for $i = 1, 2$, whence we have

$$x - \alpha_1(x_1) - \alpha_2(x_2) = (x_1 - \alpha_1(x_1)) + (x_2 - \alpha_2(x_2)) + z \in N_1' \oplus N_2'.$$

Since $\alpha_i(x_i) \in M$, we have $x \in M \oplus (N_1' \oplus N_2')$.

(2) $\Rightarrow$ (1). We use the induction on n to prove that M has the n-exchange property for every positive integer $n \geq 2$. For $n = 2$, the assertion follows from the assumption. Assume that M has the n-exchange property. We prove that M has the $(n + 1)$-exchange property.

Let $X = M \oplus Y = D_1 \oplus \cdots \oplus D_{n+1}$. We set $E = D_1 \oplus \cdots \oplus D_n$. Then $X = M \oplus Y = E \oplus D_{n+1}$. Since M has the 2-exchange property, there are two submodules $E' \subseteq E$ and $D'_{n+1} \subseteq D_{n+1}$ with $X = M \oplus E' \oplus D'_{n+1}$. We set $E'' = E \cap (M \oplus D'_{n+1})$ and $D''_{n+1} = D_{n+1} \cap (M \oplus E')$. By the modular law, $E = E' \oplus E''$ and $D_{n+1} = D'_{n+1} \oplus D''_{n+1}$. Since

$$X = M \oplus (E' \oplus D'_{n+1}) = (E'' \oplus D''_{n+1}) \oplus (E' \oplus D'_{n+1}),$$

the module E'' is isomorphic to a direct summand of M. By Theorem 1.99, E'' has the n-exchange property. In addition, $E = E' \oplus E'' = D_1 \oplus \cdots \oplus D_n$. Therefore, there are submodules $D'_i \subseteq D_i$ $(i = 1, \ldots, n)$ such that $E = E'' \oplus D'_1 \oplus \cdots \oplus D'_n$. We set $E''' = (M \oplus D'_{n+1}) \cap (E' \oplus D_{n+1})$. It follows from $E'' \subseteq M \oplus D'_{n+1} \subseteq E'' \oplus (E' \oplus D_{n+1})$ and the modular law that $M \oplus D'_{n+1} = E'' \oplus E'''$. Therefore,

$$X = E' \oplus E'' \oplus E''' = E \oplus E''' = D'_1 \oplus \cdots \oplus D'_n \oplus E'' \oplus E'''$$
$$= M \oplus D'_1 \oplus \cdots \oplus D'_{n+1}.$$

Consequently, M has the $(n+1)$-exchange property.   □

Theorem 1.108 *For a ring R, the following are equivalent.*

1. *For any two elements a and b of R with $a + b = 1$, there are two idempotents $e \in aR$ and $f \in bR$ with $e + f = 1$.*
2. *For any two elements a and b of R with $a + b = 1$, there are two idempotents $e' \in Ra$ and $f' \in Rb$ with $e' + f' = 1$.*
3. *R_R is an exchange module.*
4. *$_R R$ is an exchange module.*
5. *R_R is isomorphic to a direct summand of a finite direct sum of modules with the 2-exchange property.*
6. *$_R R$ is isomorphic to a direct summand of a finite direct sum of modules with the 2-exchange property.*
7. *Every finitely generated projective right R-module is an exchange module.*
8. *Every finitely generated projective left R-module is an exchange module.*

Proof The equivalence of conditions (1), (2), (5) and (6) follows from Theorem 1.107 and natural ring isomorphisms $\mathrm{End}(R_R) \to R$ and $\mathrm{End}\ _R R \to R$.

The implications $(3) \Rightarrow (5)$, $(7) \Rightarrow (5)$, $(4) \Rightarrow (6)$ and $(8) \Rightarrow (6)$ are obvious. The implications $(3) \Rightarrow (7)$ and $(4) \Rightarrow (8)$ follow from Theorem 1.102.

$(5) \Rightarrow (3)$. By Theorem 1.107, R_R is a module with the finite exchange property. By Theorem 1.102, the finitely generated module R_R is an exchange module.

$(6) \Rightarrow (4)$. The proof is similar to the proof of the implication $(5) \Rightarrow (3)$.

□

A ring R is called an **exchange ring** if any of the preceding equivalent conditions hold for R.

Corollary 1.109 *Let M be a nonzero module, and let $R = \mathrm{End}(M)$. Then the following conditions are equivalent.*

1. *M is a module with the finite exchange property.*
2. *R is an exchange ring.*
3. *There is a direct decomposition $M = \oplus_{i=1}^{n} M_i$ such that all the rings $\mathrm{End}(M_i)$ are exchange rings.*
4. *There is a complete set $\{e_i\}_{i=1}^{n}$ of nonzero orthogonal idempotents of the ring R such that all the rings $e_i R e_i$ are exchange rings.*
5. *For any elements $f_1, \ldots, f_n$ of R such that $\sum_{i=1}^{n} f_i = 1_M$, there are orthogonal idempotents $e_i \in R f_i$ with $\sum_{i=1}^{n} e_i = 1_M$.*
6. *For any elements $f_1, \ldots, f_n$ of R such that $\sum_{i=1}^{n} f_i = 1_M$, there are orthogonal idempotents $e_i \in f_i R$ with $\sum_{i=1}^{n} e_i = 1_M$.*

Proof The equivalence of conditions (1) and (5) follows from Theorem 1.105.

The implication $(1) \Rightarrow (2)$ follows from Theorems 1.107 and 1.108.

The implication $(2) \Rightarrow (3)$ is obvious.

$(3) \Rightarrow (4)$. Let $e_i : M \to M_i$ be the natural projections. Since there are the natural ring isomorphisms $\mathrm{End}(M_i) \to e_i R e_i$, all the rings $e_i R e_i$ are exchange rings.

$(4) \Rightarrow (3)$. We set $M_i = e_i(M)$. Then $M = \oplus_{i=1}^{n} M_i$, and all the rings $\mathrm{End}(M_i)$ are exchange rings.

$(3) \Rightarrow (1)$ By Theorem 1.107, all the modules M_i have the finite exchange property, and M is a module with the finite exchange property.

$(1) \Leftrightarrow (6)$. Since (2) is a left-right symmetric ring condition and (1) and (2) are equivalent, the equivalence of (1) and (6) follows from the equivalence of (1) and (5).

□

Corollary 1.110 *For a ring R, the following conditions are equivalent.*

1. *R is an exchange ring.*
2. *R is isomorphic to a factor ring of a direct product of exchange rings.*
3. *There is a complete set $\{e_i\}_{i=1}^{n}$ of nonzero orthogonal idempotents of R such that all the rings $e_i R e_i$ are exchange rings.*
4. *For every element a of R, there is a unitary subring S of R such that S is an exchange ring containing a.*

5. *For any elements $f_1, \ldots, f_n$ of R with $\sum_{i=1}^{n} f_i = 1$, there are orthogonal idempotents $e_i \in Rf_i$ with $\sum_{i=1}^{n} e_i = 1$.*
6. *For any elements $f_1, \ldots, f_n$ of R with $\sum_{i=1}^{n} f_i = 1$, there are orthogonal idempotents $e_i \in f_i R$ with $\sum_{i=1}^{n} e_i = 1$.*
7. *For any left ideals $B_1, \ldots, B_n$ of R with $\sum_{i=1}^{n} B_i = R$, there are orthogonal idempotents $e_i \in B_i$ with $\sum_{i=1}^{n} e_i = 1$.*
8. *For any right ideals $B_1, \ldots, B_n$ of R with $\sum_{i=1}^{n} B_i = R$, there are orthogonal idempotents $e_i \in B_i$ with $\sum_{i=1}^{n} e_i = 1$.*

Proof The equivalence of (1), (5) and (6) follows from Corollary 1.109.

The equivalences (5) $\Leftrightarrow$ (7) and (6) $\Leftrightarrow$ (8) are directly verified.

The implications (1) $\Rightarrow$ (2), (1) $\Rightarrow$ (3) and (1) $\Rightarrow$ (4) are obvious.

The implication (2) $\Rightarrow$ (1) follows from Theorem 1.108.

The implication (3) $\Rightarrow$ (1) follows from Corollary 1.109(b) and the natural ring isomorphism End $R_R \to R$.

(4) $\Rightarrow$ (1). Let a be an element of R. By assumption, there is a unitary subring S of R containing a. Since S is an exchange ring, It follows from Theorem 1.108 that there is an idempotent $e \in aS \subseteq aR$ with $1 - e \in (1-a)S \subseteq (1-a)R$. By Theorem 1.108, R is an exchange ring. $\square$

Definition 1.111 For a ring R and a right or left ideal I of R, we say that every idempotent of R can be lifted modulo I if for every element $a \in R$ with $a - a^2 \in I$, there is an idempotent $e \in R$ with $e - a \in I$. If I is an ideal of R, then we say that every idempotent of the factor ring R/I can be lifted to an idempotent of R if for every idempotent $\bar{e} \in R/I$, there is an idempotent $e \in R$ with $\bar{e} = h(e)$, where $h: R \to R/I$ is the natural epimorphism.

If I is an ideal of R, then it is easy to check that every idempotent of the factor ring R/I can be lifted to an idempotent of R if and only if every idempotent of R can be lifted modulo I.

Theorem 1.112 *For a ring R, the following conditions are equivalent.*

1. *R is an exchange ring.*
2. *For every element $x \in R$, there is an idempotent $e \in R$ with $e - x \in (x - x^2)R$.*
3. *Every idempotent of R can be lifted modulo every right ideal.*
4. *$R/J(R)$ is an exchange ring and every idempotent of $R/J(R)$ can be lifted to an idempotent of R.*
5. *There is an ideal I of R such that $I \subseteq J(R)$, R/I is an exchange ring and every idempotent of R/I can be lifted to an idempotent of R.*

6. For every element x of R, there is an idempotent $e \in xR$ with
   $R = eR + (1 - x)R$.
7. For every idempotent e of R, the ring eRe is an exchange ring.

Proof $(1) \Rightarrow (2)$. Since R is an exchange ring, there is an idempotent $e \in xR$ such that $1 - e \in (1 - x)R$. Then $e - x = (1 - x)e - x(1 - e) \in (1 - x)xR + x(1 - x)R = (x - x^2)R$.

$(2) \Rightarrow (3)$. Let M be a right ideal of R, and let x be an element of R with $x - x^2 \in M$. By (2), there is an idempotent $e \in R$ such that $e - x \in (x - x^2)R \subseteq M$.

$(3) \Rightarrow (1)$. Let $x \in R$. Since $x - x^2 \in (x - x^2)R$ and every idempotent can be lifted modulo the right ideal $(x - x^2)R$, there is an idempotent $e \in R$ such that $e - x \in (x - x^2)R \subseteq xR \cap (1 - x)R$. Then $e = (e - x) + x \in xR$ and $1 - e = (1 - x) - (e - x) \in (1 - x)R$.

The implications $(1) + (3) \Rightarrow (4)$, $(4) \Rightarrow (5)$ and $(6) \Rightarrow (1)$ are directly verified.

$(5) \Rightarrow (6)$. Let $x \in R$, and let $h: R \to R/I$ be the natural epimorphism. Since $h(R)$ is an exchange ring, there are two elements $a, c \in R$ such that $a^2 - a \in xR + I$ and $1 - a - (1 - x)c \in I$. Without loss of generality, we can assume that $a \in xR$. Since $h(a)$ is an idempotent of $h(R)$, it follows from (5) that there is an idempotent f of R such that $f - a \in I \subseteq J(R)$. Then $1 - f + a = u$ is an invertible element of R, $uf = af \in xR$, and $1 - u = f - a \in I$, whence $1 - u^{-1} \in I$. We set $e = ufu^{-1} = afu^{-1}$. Then $e^2 = e \in xR$ and $f - e = f(1 - u^{-1}) + (1 - u)fu^{-1} \in I$. Therefore, $(1 - e) - (1 - x)c = (a - f) + (f - e) + (1 - a) - (1 - x)c \in I \subseteq J(R)$, whence $(1 - e)R \subseteq (1 - x)R + I$. Then $R = eR + (1 - e)R = eR + (1 - x)R + I$. Since $I \subseteq J(R)$, we have $R = eR + (1 - x)R$.

$(6) \Rightarrow (1)$. Let $x \in R$. By (6), there is an idempotent $e \in xR$ such that $R = eR + (1 - x)R$. Then there is an element $t \in R$ such that $1 - et \in (1 - x)R$. Let $f = e + et(1 - e)$. Then $f^2 = f \in eR \subseteq xR$ and $1 - f = (1 - et) - (1 - et)e \in (1 - x)R$. Therefore, R is an exchange ring.

$(1) \Rightarrow (6)$. Since eR is a direct summand of R_R, it follows from Theorem 1.99 that eR is a module with the finite exchange property. Since $eRe \cong \operatorname{End}(eR)$, eRe is an exchange ring by Corollary 1.109(b). $\square$

Proposition 1.113 *A ring R is an exchange ring without nontrivial idempotents if and only if R is a local ring.*

Proof If R is a local ring, then it follows from Theorem 1.112 (see condition (4)) that R is an exchange ring.

Let R be an exchange ring without nontrivial idempotents, let $a, b \in R$, and let $a + b = 1$. By Theorem 1.107, there are $e = e^2 \in aR$ and $f = f^2 \in bR$ with $e + f = 1$. By assumption, every nonzero idempotent of R is equal to 1. Therefore, either $e = 1$ or $f = 1$. Then either $aR = R$ or $bR = R$. Therefore, R is a local ring. $\qquad\square$

Theorem 1.114 *For a module M, the following conditions are equivalent.*

1. *M is an indecomposable exchange module.*
2. *M is an indecomposable module with the 2-exchange property.*
3. *End M is a local ring.*

Proof We set $R = \mathrm{End}(M)$.

$(1) \Rightarrow (3)$ Since M is an indecomposable module, the ring R does not have nontrivial idempotents. Since M is an exchange module, R is an exchange ring by Corollary 1.109. By Proposition 1.113, R is a local ring.

$(3) \Rightarrow (2)$ By Proposition 1.113, R is an exchange ring. By Corollary 1.109, M is a module with the 2-exchange property. The local ring R does not have nontrivial idempotents. Therefore, M is an indecomposable module.

$(2) \Rightarrow (1)$ The module M with the 2-exchange property is a module with the finite exchange property by Theorem 1.107. Let $X = M \oplus N = \oplus_{i \in I} Y_i$. Since every element of X is contained in a sum of finitely many modules Y_i, there is a finite subset J of the set I with $M \cap (\oplus_{i \in J} Y_i) \neq 0$. We set $E = \oplus_{i \in I \setminus J} Y_i$. Then $X = (\oplus_{i \in J} Y_i) \oplus E$. Since M is a module with the finite exchange property, there are submodules $E' \subseteq E$ and $Y_i' \subseteq Y_i$ ($i \in J$) with $X = M \oplus (\oplus_{i \in J} Y_i') \oplus E'$. Since E contains the direct summand E' of X and Y_i contains the direct summand Y_i' of X, $i \in J$, there are direct decompositions $E = E' \oplus E''$ and $Y_i = Y_i' \oplus Y_i''$ ($i \in J$). Since

$$X = (\oplus_{i \in J} Y_i) \oplus E = (\oplus_{i \in J} Y_i') \oplus E' \oplus (\oplus_{i \in J} Y_i'') \oplus E''$$
$$= M \oplus (\oplus_{i \in J} Y_i') \oplus E',$$

the indecomposable module M is isomorphic to the module $(\oplus_{i \in J} Y_i'') \oplus E''$. Since $M \cap (\oplus_{i \in J} Y_i) \neq 0$ and $M \cap (\oplus_{i \in J} Y_i') = 0$, we have $\oplus_{i \in J} Y_i'' \neq 0$. In addition, the module M is indecomposable, so $E'' = 0$. Then $E' = E$, and we have

$$X = M \oplus (\oplus_{i \in J} Y_i') \oplus E = M \oplus (\oplus_{i \in J} Y_i') \oplus (\oplus_{i \in I \setminus J} Y_i).$$

Therefore, M is an exchange module. $\qquad\square$

Theorem 1.115 *For a nonzero module M, the following conditions are equivalent.*

1. *M is an exchange module that is a direct sum of finitely many indecomposable modules.*
2. *M is a direct sum of finitely many indecomposable modules with the 2-exchange property.*
3. *M is a direct sum of finitely many modules whose endomorphism rings are local rings.*
4. *M is a direct sum of finitely many modules whose endomorphism rings are semiperfect rings.*
5. End(M) *is a semiperfect ring.*

Proof We set $R = \text{End}(M)$. The implication $(1) \Rightarrow (2)$ follows from Theorem 1.101.

The implication $(2) \Rightarrow (3)$ follows from Theorem 1.113.

$(3) \Rightarrow (4)$. By assumption, there is a direct decomposition $M = \oplus_{i=1}^{n} M_i$ such that all the rings End(M_i) are local rings. By Theorem 1.113, all the modules M_i are indecomposable exchange modules. By Theorems 1.107 and 1.108, all the rings End(M_i) are orthogonally finite exchange rings.

$(4) \Rightarrow (2)$ By assumption, there is a direct decomposition $M = \oplus_{i=1}^{n} M_i$ such that all the rings End(M_i) are orthogonally finite exchange rings. By Corollary 1.109, all the modules M_i are modules with the finite exchange property. By Theorems 1.107 and 1.108, M is a module with the finite exchange property. Since all the rings End(M_i) are orthogonally finite, every module M_i is a direct sum of finitely many indecomposable modules. Therefore, M is the direct sum of finitely many indecomposable modules $X_1, \ldots, X_k$. By Theorem 1.99, every direct summand X_i of M is a module with the finite exchange property.

$(2) \Rightarrow (5)$. By assumption, there is a complete set $\{e_i\}_{i=1}^{n}$ of orthogonal nonzero idempotents of the ring R such that all the modules $e_i(M)$ are indecomposable modules with the 2-exchange property. It follows from Theorem 1.113 that all the idempotents e_i of R are local idempotents. By Definition 1.36, R is a semiperfect ring.

$(5) \Rightarrow (3)$. By Definition 1.36, there is a complete set $\{e_i\}_{i=1}^{n}$ of nonzero orthogonal idempotents of R such that all the rings $e_i R e_i$ are local rings. If we set $M_i = e_i(M)$, then $M = \oplus_{i=1}^{n} M_i$ and all the rings End(M_i) are local rings.

$(3) \Rightarrow (1)$. By Theorem 1.113, all the modules M_i are indecomposable exchange modules. By Theorem 1.101, M is an exchange module. $\square$

Definition 1.116 A ring R is called a clean ring if each element $a \in R$ can be expressed as $a = e + u$, where e is an idempotent in R and u is a unit in R.

It is not difficult to see that clean rings are exchange rings but the converse is not always true. It is known that if R is an exchange ring with central idempotents, then R is a clean ring. A semiperfect ring is always a clean ring. If R is a clean ring and R does not contain an infinite set of orthogonal idempotents, then R is semiperfect [84].

Definition 1.117 A module M is called a clean module if $\text{End}(M)$ is a clean ring.

It was shown in [20] that the class of clean modules includes continuous modules, discrete modules, flat cotorsion modules and quasi-projective right modules over a right perfect ring.

Utumi gave structure of endomorphism rings of quasi-injective modules in [129] by proving the following.

Theorem 1.118 *Let M be a quasi-injective right R-module, and let $S = \text{End}(M)$. Then $J(S) = \{f \in S \colon \text{Ker}(f) \subseteq_e M\}$, and $S/J(S)$ is a von Neumann regular right self-injective ring whose idempotents lift modulo $J(S)$.*

1.11 Pure-Injective and Cotorsion Modules

Definition 1.119 A short exact sequence

$$0 \to N \to M \to M/N \to 0$$

in Mod-R is called *pure* if the induced sequence

$$0 \to N \otimes_R X \to M \otimes_R X \to (M/N) \otimes_R X \to 0$$

remains exact in Ab for any left R-module X and, equivalently, if any finitely presented module F is projective with respect to it.

Definition 1.120 A module $E \in$ Mod-R is called *pure-injective* if it is injective with respect to any pure-exact sequence.

When dealing with pure-injectivity, the so-called functor category technique is quite useful. Let us briefly explain the basic ideas behind this technique. An abelian category $\mathcal{C}$ is called a *Grothendieck category* if $\mathcal{C}$ has coproducts, exact direct limits and a generator set of objects. And, a Grothendieck category $\mathcal{C}$ is called *locally finitely presented* if it has a generator set $\{C_i\}_{i \in \mathcal{I}}$ consisting of finitely presented objects. Recall that an object $C \in \mathcal{C}$ is called finitely presented if the functor $\text{Hom}_{\mathcal{C}}(C, -) \colon \mathcal{C} \to Ab$ commutes with direct limits. Every locally finitely presented Grothendieck category $\mathcal{C}$ has enough injective objects, and every object $C \in \mathcal{C}$ can be essentially embedded in an injective object $E(C)$, called the injective envelope of C.

It is well known (see e.g. [17, 88]) that one can associate to any module category Mod-R, a locally finitely presented Grothendieck category $\mathcal{C}$, which is usually called the functor category of Mod-R, and a fully faithful embedding

$$F : \text{Mod} - R \to \mathcal{C}$$

satisfying the following properties:

1. The functor F has a right adjoint functor $G : \mathcal{C} \to \text{Mod-}R$.
2. A short exact sequence

$$\Sigma \equiv 0 \to X \to Y \to Z \to 0$$

   in Mod-R is pure if and only if the induced sequence $F(\Sigma)$ is exact (and pure) in $\mathcal{C}$.
3. F preserves finitely generated objects; i.e. the image of any finitely generated object in Mod-R is a finitely generated object in $\mathcal{C}$.
4. F identifies Mod-R with the full subcategory of $\mathcal{C}$ consisting of the all FP-injective objects in $\mathcal{C}$ where an object $C \in \mathcal{C}$ is FP-injective if $\text{Ext}^1_{\mathcal{C}}(X, C) = 0$ for every finitely presented object $X \in \mathcal{C}$.
5. A module $M \in$ Mod-R is pure-injective if and only if $F(M)$ is an injective object of $\mathcal{C}$.
6. Every module $M \in$ Mod-R admits a pure embedding in a pure-injective object $PE(M) \in$ Mod-R such that the image of this embedding under F is the injective envelope of $F(M)$ in $\mathcal{C}$. The pure-injective object $PE(M)$ is called the *pure-injective envelope* of M.

As a consequence, we have the following useful observation:

Proposition 1.121 *Let M be a right R-module. Then M is invariant under automorphisms (resp., endomorphisms) of its pure-injective envelope in Mod-R if and only if $F(M)$ is invariant under automorphisms (resp., endomorphisms) of its injective envelope in $\mathcal{C}$.*

Definition 1.122 A module M in Mod-R is called quasi pure-injective if it is invariant under endomorphisms of its pure-injective envelope and, equivalently, if $F(M)$ is a quasi-injective object in the associated functor category $\mathcal{C}$.

It is well known that any object C in a Grothendieck category admits a minimal embedding $u : C \to Q$ in a quasi-injective object Q, which is called its quasi-injective envelope. In particular, this shows that, for any module M, the object $F(M)$ has a quasi-injective envelope $u : F(M) \to Q$ in the functor category $\mathcal{C}$ of Mod-R. By construction, $F(M)$ is an FP-injective object of $\mathcal{C}$, but we cannot see any reason why this object Q must be also FP-injective and, thus, belong to the image of the functor F. As a consequence, it seems that, in general, modules in Mod-R do not need to have a quasi pure-injective envelope.

The next example sheds some light to the possible consequences that this lack of quasi pure-injective envelopes may have in the characterization of quasi pure-injective modules.

Example 1.123 Let R be a commutative PID. Then an R-module M is flat if and only if it is torsion-free, and therefore, the only possible pure ideals of R are 0 and R itself. This means that any homomorphism $f : N \to R$ from a pure ideal N of R to R trivially extends to an endomorphism of R. However, if R would always be quasi pure-injective, then we would get that

$$R \cong \operatorname{End}_R(R) \cong \operatorname{End}_{\mathcal{C}}(F(R)) \cong \operatorname{End}_{\mathcal{C}}(E(F(R))),$$

where $E(F(R))$ is the injective envelope of $F(R)$ in $\mathcal{C}$. And, as the endomorphism ring of an injective object in a Grothendieck category is always left pure-injective, this would mean that any commutative PID is pure-injective. But this is not the case, as, for instance, the case of $\mathbb{Z}$ shows.

Definition 1.124 A pair $(\mathcal{F}, \mathcal{C})$ is called a cotorsion pair if $\mathcal{F}, \mathcal{C}$ are two classes of modules such that

1. $F \in \mathcal{F} \Leftrightarrow \operatorname{Ext}^1(F, C) = 0$ for all $C \in \mathcal{C}$.
2. $C \in \mathcal{C} \Leftrightarrow \operatorname{Ext}^1(F, C) = 0$ for all $F \in \mathcal{F}$.

Let us assume that $\mathcal{F}$ is closed under direct limits and that any module has an $\mathcal{F}$-cover (and therefore a $\mathcal{C}$-envelope). This is true, for instance, if there exists a subset $\mathcal{F}_0 \subset \mathcal{F}$ such that $C \in \mathcal{C} \Leftrightarrow \operatorname{Ext}^1(F, C) = 0$ for all $F \in \mathcal{F}_0$ (see e.g. [30, theorem 2.6] or [37, chapter 6]). It is well known that in a cotorsion pair, any $\mathcal{C}$-envelope $u : M \to C(M)$ is monomorphic and any $\mathcal{F}$-cover $p : F(M) \to M$ is epimorphic.

In the particular case in which $\mathcal{F}$ is the class of flat modules, the cotorsion pair $(\mathcal{F}, \mathcal{C})$ is usually called the *Enochs cotorsion pair* [37, definition 5.18, p. 122], and modules in $\mathcal{C}$ are just called *cotorsion modules*. By the definition of a cotorsion pair, we get that a module M is cotorsion if and only if $\operatorname{Ext}^1(F, M) = 0$ for every flat module. A ring R is called *right cotorsion* if R_R is a cotorsion module.

Lemma 1.125 *Assume that* $(\mathcal{F}, \mathcal{C})$ *is a cotorsion pair, and* $\mathcal{F}$ *is closed under direct limits. Let* $X \in \mathcal{F} \cap \mathcal{C}$. *Then* $S = \operatorname{End}(X)$ *is a right cotorsion ring.*

Proof We first wish to show that $S = \operatorname{End}(X)$ is a right cotorsion ring. Let F be a flat right S-module, and let

$$\Sigma: \quad 0 \to S \xrightarrow{u} N \xrightarrow{p} F \to 0$$

be any extension of F by S. We must show that Σ splits. As F is flat, we may write it as a direct limit of free modules of finite rank, say $F = \varinjlim S^{n_i}$. Now, applying the tensor functor $T = - \otimes_S X$, we get in Mod-R the sequence

$$0 \to T(S) \xrightarrow{T(u)} T(N) \xrightarrow{T(p)} T(F) \to 0,$$

which is exact since F is flat. Moreover, $T(S) \cong X$.

On the other hand, as $T = - \otimes_S X$ commutes with direct limits, we get that

$$T(F) = T(\varinjlim S^{n_i}) \cong \varinjlim T(S^{n_i}) \cong \varinjlim X^{n_i}$$

and, thus, $T(F)$ belongs to $\mathcal{F}$ since we are assuming that $\mathcal{F}$ is closed under direct limits. As $X_R \in \mathcal{C}$, we get that the preceding sequence splits, and therefore, there exists a $\pi: T(N) \to T(S)$ such that $\pi \circ T(u) = 1_{T(S)}$. Now, applying the functor $H = \operatorname{Hom}_R(X, -)$, we get a commutative diagram in Mod-S

$$
\begin{array}{ccccccccc}
0 & \to & S & \xrightarrow{u} & N & \xrightarrow{p} & F & \to & 0 \\
 & & \downarrow{\sigma_S} & & \downarrow{\sigma_N} & & \downarrow{\sigma_F} & & \\
0 & \to & HT(S) & \xrightarrow{HT(u)} & HT(N) & \xrightarrow{HT(p)} & HT(F) & &
\end{array}
$$

in which $\sigma: 1_{\text{Mod}-S} \to HT$ is the arrow of the adjunction and σ_S is an isomorphism. Then we have that

$$\sigma_S^{-1} \circ H(\pi) \circ \sigma_N \circ u = \sigma_S^{-1} \circ H(\pi) \circ HT(u) \circ \sigma_S$$

$$= \sigma_S^{-1} \circ 1_{HT(S)} \circ \sigma_S = 1_S,$$

and this shows that Σ splits.

Therefore, $S = \operatorname{End}(X)$ is right cotorsion ring. $\qquad\square$

Recall that two ideals A and B in a commutative ring R are called *coprime* when $A + B = R$ – that is, when there exist elements $a \in A$ and $b \in B$ satisfying the Bezout identity. Motivated by it, the following definition was given in [44].

Definition 1.126 Two elements a, b of a ring R form a left *coprime pair* if there exist elements $r, s \in R$ such that $ra + sb = 1$ and, equivalently, if $R = Ra + Rb$.

One can define an order relation among left coprime pairs as follows: $(a, b) \leq (a', b')$ if and only if $Ra \subseteq Ra'$ and $Rb \subseteq Rb'$.

Remarks 1.127 1. There is some abuse of notation in this definition of order relation since it does not need to satisfy the antisymmetric property. This is easily solved by defining an equivalence relation in R by $(a, b) \sim (a', b')$ if and only if $Ra = Ra'$ and $Rb = Rb'$.
2. A pair (a, b) is left coprime if the sum of the cyclic left ideals generated by its elements is the whole ring. Thus, this order relation reflects how far is R from being the direct sum of those ideals. Therefore, this order relation is especially useful to construct idempotents in the ring.
3. Given an element $a \in R$, we may associate to it the left coprime pair $(a, 1 - a)$. We may then define the following order relation in R: $a \leq b$ if and only if $(a, 1 - a) \leq (b, 1 - b)$. Moreover, note that, given a left coprime pair (a, b), there exist elements $r, s \in R$ such that $1 = ra + sb$. Therefore, calling $c = ra$, we get that $(c, 1 - c) \leq (a, b)$. In other words, *for every left coprime pair (a,b), there exists an element $c \in R$ such that $c \leq (a, b)$*. In particular, this means that every minimal left coprime pair (a, b) is equivalent to one of the form $(c, 1 - c)$ for some $c \in R$.

The next proposition shows that this order relation describes idempotents in rings.

Proposition 1.128 *[44, proposition 3] An element $a \in R$ is idempotent if and only if $(a, 1 - a)$ is a minimal left coprime pair.*

We can now reformulate the characterization of exchange rings given in Theorem 1.112 as follows.

Theorem 1.129 *Let R be a ring. The following are equivalent.*

1. *R is an exchange ring.*
2. *For every $a \in R$, there exists an idempotent $e \in R$ such that $e \leq a$.*
3. *For every $a \in R$, there exists a minimal element $b \in R$ such that $b \leq a$.*

The reason why left coprime pairs are useful in the study of right cotorsion rings is the following result that was proved in [44].

Theorem 1.130 *Let R be a right cotorsion ring and (a, b), a left coprime pair. Then there exists a minimal left coprime pair (a', b') below it. This minimal coprime pair can be chosen of the form $(e, 1 - e)$ with e an idempotent of R.*

Corollary 1.131 *Let R be a right cotorsion ring. Then,*

1. *R is an exchange ring.*
2. *An element $a \in R$ belongs to the Jacobson radical if and only if $0 \leq a$. In particular, a left ideal I is contained in $J(R)$ if and only if it contains no nonzero idempotents.*

Proof The first assertion is an immediate consequence of Theorems 1.129 and 1.130. For the second assertion, let us first note that, if $a \in J(R)$ and $e \in R$ is an idempotent satisfying that $e \leq a$, then e also belongs to $J(R)$ and, thus, $e = 0$. On the other hand, if $a \notin J(R)$, then there exists an $r \in R$ such that $1 - ar$ is not invertible. As R is an exchange ring, there exists an idempotent $e \leq a$. And $e \neq 0$ since, otherwise, $1 - ar$ would be a unit. $\square$

Indeed, Guil Asensio and Herzog proved in [44] a stronger structure theorem for right cotorsion rings.

Theorem 1.132 *If R is a right cotorsion ring, then $R/J(R)$ is a von Neumann regular, right self-injective ring and idempotents lift modulo any two-sided ideal containing $J(R)$.*

Corollary 1.133 *Let R be a right perfect ring, and let P be a projective right R-module. Then $\mathrm{End}_R(P)/J(\mathrm{End}_R(P))$ is a von Neumann regular, right self-injective ring and idempotents lift modulo any two-sided ideal containing $J(\mathrm{End}_R(P))$.*

Proof As R is right perfect, every flat right R-module is projective, and thus, every right R-module is cotorsion. In particular, P is a flat cotorsion right R-module. Thus, $\mathrm{End}(P)$ is a right cotorsion ring by Lemma 1.125. The result now follows from the preceding theorem. $\square$

2

Modules Invariant under Automorphisms of Envelopes

Dickson and Fuller in [22] initiated the study of modules invariant under automorphisms of their injective envelopes, but the investigation of the general theory of modules invariant under automorphisms of their envelopes and covers was initiated by Guil Asensio, Tütüncü and Srivastava in [49] and was continued further in [50] and [53]. In this chapter, we discuss the basics of the general theory of envelopes and then study the modules invariant under automorphisms of their envelopes.

2.1 Introduction to Envelopes

Eckmann and Schopf proved the existence of injective envelope for any module [25]. Warfield [132] proved the existence of the pure-injective envelope. Bass introduced the dual notion of projective cover of a module over a right perfect ring [9]. Motivated by the notions of injective envelope and projective cover, Auslander and Enochs independently introduced the general notion of envelopes and covers.

Let R be any ring. Let us fix a nonempty class of right R-modules $\mathcal{X}$, closed under isomorphisms and direct summands.

We first define the notions of $\mathcal{X}$-preenvelope and $\mathcal{X}$-envelope.

Definition 2.1 An $\mathcal{X}$-preenvelope of a right module M is a homomorphism $u \colon M \to X(M)$ with $X(M) \in \mathcal{X}$ such that any other homomorphism $u' \colon M \to X'$ with $X' \in \mathcal{X}$ factors through u; that is, there exists a homomorphism $f \colon X(M) \to X'$ with $f \circ u = u'$.

$$M \xrightarrow{\ \ u\ \ } X(M)$$

$$\downarrow{\scriptstyle u'} \quad {\nearrow}$$

$$X' \quad {}_{\kappa}{\nearrow}{\scriptstyle f}$$

Definition 2.2 A preenvelope $u: M \rightarrow X(M)$ is called an $\mathcal{X}$-envelope if, whenever there is an endomorphism $h: X(M) \rightarrow X(M)$ such that $h \circ u = u$, then h is an automorphism.

It is clear from the definition that the $\mathcal{X}$-envelope of a module M is unique up to isomorphism. This means that if $u_1: M \rightarrow X_1(M)$ and $u_2: M \rightarrow X_2(M)$ are two $\mathcal{X}$-envelopes, then $X_1(M) \cong X_2(M)$.

Definition 2.3 The class $\mathcal{X}$ is called an enveloping class if each right R-module M has an $\mathcal{X}$-envelope.

Definition 2.4 An $\mathcal{X}$-(pre)envelope $u: M \rightarrow X(M)$ is called monomorphic if u is a monomorphism.

Examples 2.5 Here are some examples of enveloping classes.

1. Let $\mathcal{X}$ be the class of injective right R-modules. Then $\mathcal{X}$ is an enveloping class, and in this case, the $\mathcal{X}$-envelope $u: M \rightarrow X(M)$ is a monomorphic $\mathcal{X}$-envelope. This $\mathcal{X}$-envelope $u: M \rightarrow X(M)$ is the same as the injective envelope defined by Eckmann and Schopf. Here $X(M)$ is usually denoted as $E(M)$.
2. Let $\mathcal{X}$ be the class of pure-injective right R-modules. Then $\mathcal{X}$ is an enveloping class. In this case, too, the $\mathcal{X}$-envelope $u: M \rightarrow X(M)$ is a monomorphic $\mathcal{X}$-envelope, and it is same as the pure-injective envelope defined by Warfield. Here $X(M)$ is usually denoted as $PE(M)$.
3. Let $\mathcal{X}$ be the class of cotorsion right R-modules. Then $\mathcal{X}$ is an enveloping class. In this case, too, the $\mathcal{X}$-envelope $u: M \rightarrow X(M)$ is a monomorphic $\mathcal{X}$-envelope, and it is same as the cotorsion envelope defined by Bican, El Bashir and Enochs [11]. Here $X(M)$ is usually denoted as $C(M)$.

Theorem 2.6 *If $u_1: M_1 \rightarrow X(M_1)$ and $u_2: M_2 \rightarrow X(M_2)$ are $\mathcal{X}$-envelopes, then $u_1 \oplus u_2: M_1 \oplus M_2 \rightarrow X(M_1) \oplus X(M_2)$ is an $\mathcal{X}$-envelope.*

Proof Let $u': M_1 \oplus M_2 \rightarrow X'$ be any homomorphism where $X' \in \mathcal{X}$. If $v_i: M_i \rightarrow M_1 \oplus M_2$ is the canonical injection for $i = 1, 2$, then $u' \circ v_i: M_i \rightarrow X'$ is a homomorphism. Now, since $u_i: M_i \rightarrow X(M_i)$ is an $\mathcal{X}$-envelope, we have a homomorphism $h_i: X(M_i) \rightarrow X'$ such that $h_i \circ u_i = u' \circ v_i$. Consider the unique homomorphism $f: X(M_1) \oplus X(M_2) \rightarrow X'$ such that $f|_{X(M_i)} = h_i$.

Then $u' = f \circ (u_1 \oplus u_2)$. This shows that $u_1 \oplus u_2 \colon M_1 \oplus M_2 \to X(M_1) \oplus X(M_2)$ is an $\mathcal{X}$-preenvelope. Next, suppose $h \colon X(M_1) \oplus X(M_2) \to X(M_1) \oplus X(M_2)$ is an endomorphism such that $h \circ (u_1 \oplus u_2) = u_1 \oplus u_2$. It may be verified that h is an automorphism, thus showing that $u_1 \oplus u_2 \colon M_1 \oplus M_2 \to X(M_1) \oplus X(M_2)$ is an $\mathcal{X}$-envelope. $\qquad\square$

The preceding argument can be easily extended to any finite direct sum.

2.2 Modules Invariant under Endomorphisms and Automorphisms

Definition 2.7 A module M with an $\mathcal{X}$-envelope $u \colon M \to X(M)$ is said to be $\mathcal{X}$-endomorphism-invariant if for any endomorphism $g \colon X(M) \to X(M)$ there exists an endomorphism $f \colon M \to M$ such that $u \circ f = g \circ u$.

$$
\begin{array}{ccc}
M & \dashrightarrow{\ \ f\ \ } & M \\
\downarrow{\scriptstyle u} & & \downarrow{\scriptstyle u} \\
X(M) & \xrightarrow{\ \ g\ \ } & X(M)
\end{array}
$$

Examples 2.8 For the particular cases, when $\mathcal{X}$ is the class of all injective modules, or pure-injective modules, we have the following.

1. If $\mathcal{X}$ is the class of injective modules, then the $\mathcal{X}$-endomorphism-invariant modules are precisely the quasi-injective modules.
2. If $\mathcal{X}$ is the class of pure-injective modules, then the $\mathcal{X}$-endomorphism-invariant modules are just the modules that are strongly invariant in their pure-injective envelopes in the sense of [61, definition, p. 430].

Definition 2.9 A module M with an $\mathcal{X}$-envelope $u \colon M \to X(M)$ is said to be $\mathcal{X}$-automorphism-invariant if for any automorphism $g \colon X(M) \to X(M)$ there exists an endomorphism $f \colon M \to M$ such that $u \circ f = g \circ u$.

It may be seen that the endomorphism f in the preceding definition turns out to be an automorphism as is proved in the next lemma.

Lemma 2.10 *Let M be an $\mathcal{X}$-automorphism-invariant module with an $\mathcal{X}$-envelope $u \colon M \to X(M)$. If $g \colon X(M) \to X(M)$ is an automorphism and $f \in \mathrm{End}(M)$ such that $u \circ f = g \circ u$, then f is an automorphism.*

Proof As g is an automorphism, there exists an $f' \in \text{End}(M)$ such that $u \circ f' = g^{-1} \circ u$, and thus, $u \circ f \circ f' = u \circ f' \circ f = u$. Therefore, $f \circ f'$ and $f' \circ f$ are automorphisms by the definition of envelope. This shows that f is an automorphism. $\qquad\qquad\qquad\qquad\qquad\qquad\qquad\qquad\qquad\qquad\square$

Remark 2.11 (1) The preceding definition of $\mathcal{X}$-automorphism-invariant module is equivalent to assert that M is invariant under the group action on X given by $\text{Aut}(X)$. Moreover, in this case, by the preceding lemma, the map

$$\Delta \colon \text{Aut}(X) \to \text{Aut}(M),$$

which assigns $g \mapsto f$, is a surjective group homomorphism whose kernel consists of those automorphisms $g \in \text{Aut}(X)$ such that $g \circ u = u$. This subgroup of $\text{Aut}(X)$ is usually called the *Galois group* of the envelope u, and we will denote it by $\text{Gal}(u)$. Therefore, we get that, for modules M having monomorphic $\mathcal{X}$-envelopes, M is $\mathcal{X}$-automorphism-invariant precisely when the envelope u induces a group isomorphism $\text{Aut}(M) \cong \text{Aut}(X)/Gal(u)$.

(2) The preceding definition of $\mathcal{X}$-automorphism-invariant modules can be easily extended to modules having $\mathcal{X}$-preenvelopes. We restrict our definition to modules having envelopes because these are the modules to which our results will be applied in practice.

The following example shows that, in general, $\mathcal{X}$-automorphism-invariant modules need not be $\mathcal{X}$-endomorphism-invariant.

Example 2.12 Let R be the ring of all eventually constant sequences $(x_n)_{n \in \mathbb{N}}$ of elements in $\mathbb{F}_2$, and $\mathcal{X}$, the class of injective right R-modules. As R is von Neumann regular, the class of injective R-modules coincides with the class of pure-injective R-modules as well as with the class of flat cotorsion R-modules. The $\mathcal{X}$-envelope of R_R is $u \colon R_R \to X$, where $X = \prod_{n \in \mathbb{N}} \mathbb{F}_2$. Clearly, X has only one automorphism, namely the identity automorphism. Thus, R_R is $\mathcal{X}$-automorphism-invariant, but it is not $\mathcal{X}$-endomorphism-invariant.

Example 2.13 Let R be a von Neumann regular ring that is not semisimple Artinian. Then there exists a right R-module A such that A is quasi-injective but not injective. Let S be a ring with a right S-module N' that is pure-injective but not injective. Let $N = E(S) \oplus N'$, $T = R \times S$ and $M_T = A \times N$. Note that the pure-injective envelope of M, $PE(M) = E(A) \times N$. Clearly, M_T is invariant under endomorphisms of $PE(M)$. Note that M is not invariant under endomorphisms of its injective envelope because if M is quasi-injective, then N would also be quasi-injective, and as $S \subset N$, this would mean that N is injective. This gives us a contradiction to the fact that N', a direct summand of N, is not injective.

Lemma 2.14 *Let M be an $\mathcal{X}$-automorphism-invariant module and $u : M \rightarrow X$, its $\mathcal{X}$-envelope. If each element of $\mathrm{End}(X)/J(\mathrm{End}(X))$ is a sum of units, then M is $\mathcal{X}$-endomorphism-invariant.*

Proof First, note that if each element in $\mathrm{End}(X)/J(\mathrm{End}(X))$ is a sum of units, then each element in $\mathrm{End}(X)$ is also a sum of units. Now, as M is $\mathcal{X}$-automorphism-invariant, we get that M is $\mathcal{X}$-endomorphism-invariant. $\quad\square$

The preceding lemma shows that the question of when an $\mathcal{X}$-endomophism-invariant module is $\mathcal{X}$-automorphism-invariant is closely connected to the problem of understanding when endomorphism rings of modules are generated additively by unit elements. We will discuss it in the next section.

2.3 Additive Unit Structure of von Neumann Regular Rings

The historical origin of study of the additive unit structure of rings may be traced back to the work of Dieudonné on Galois theory of simple and semisimple rings [23]. In [59], Hochschild studied additive unit representations of elements in simple algebras and proved that each element of a simple algebra over any field is a sum of units. Later, Zelinsky [139] proved that the ring of linear transformations is generated additively by its unit elements. Zelinsky showed that every linear transformation of a vector space V over a division ring is a sum of two invertible linear transformations, except when V is one-dimensional over $\mathbb{F}_2$, the field of two elements. Zelinsky also noted in his paper that this result follows from a previous result of Wolfson [135].

Apart from the ring of linear transformations, there are several other natural classes of rings that are generated by their unit elements. Let X be a completely regular Hausdorff space. Then every element in the ring $C(X)$ of real-valued continuous functions on X is the sum of two units. For any $f(x) \in C(X)$, we have $f(x) = [(f(x) \vee 0) + 1] + [(f(x) \wedge 0) - 1]$. Every element in a real or complex Banach algebra B is the sum of two units. For any $a \in B$, there exists a scalar λ ($\neq 0$) such that $a - \lambda$ is a unit by spectral radius theorem. Now $a = (a - \lambda) + \lambda$ clearly shows that a is the sum of two units.

On the other hand, any ring having a homomorphic image isomorphic to $\mathbb{F}_2 \times \mathbb{F}_2$ cannot be additively generated by its units because in $\mathbb{F}_2 \times \mathbb{F}_2$, the element $(1, 0)$ cannot be expressed as a sum of any number of units. Skornyakov [95] asked if R is a von Neumann regular ring that does not have a homomorphic image isomorphic to $\mathbb{F}_2 \times \mathbb{F}_2$, is R additively generated by its units? Skornyakov's question was answered in the negative by George Bergman. Bergman's example given in what follows was first reported in a paper by Handelman [56]. This gives an example of a von Neumann regular ring in which not all elements are sums of units.

Bergman's Example. Let k be any field, and $A = k[[x]]$ be the power series ring in one variable over k. Let K be the field of fractions of A. All ideals of A are generated by a power of x; we denote these ideals by (x^n). Let $R = \{r \in \operatorname{End}(A_k): \text{there exists } q \in K, \text{a positive integer } n, \text{with } r(a) = qa \text{ for all } a \in (x^n)\}$. Then R is a von Neumann regular ring that is not additively generated by its units.

Type theory of von Neumann regular right self-injective rings. Kaplansky proposed a classification for Baer rings [72]. Since von Neumann regular right self-injective rings are Baer rings, this classification theory of Baer rings applies to them. A von Neumann regular ring is called abelian regular if all its idempotents are central. A ring R is called a directly finite ring if for each $x, y \in R$, $xy = 1$ implies $yx = 1$. A module M is called a directly finite module if M is not isomorphic to a proper direct summand of itself. It turns out that a module M is a directly finite module if and only if $\operatorname{End}(M)$ is a directly finite ring. A ring R is called a purely infinite ring if $R_R \cong R_R \oplus R_R$. An idempotent e in a von Neumann regular right self-injective ring R is called an abelian idempotent if the ring eRe is an abelian regular ring. An idempotent e in a von Neumann regular right self-injective ring is called faithful idempotent if 0 is the only central idempotent orthogonal to e; that is, $ef = 0$ implies $f = 0$, where f is a central idempotent. A von Neumann regular right self-injective ring is said to be of Type I provided it contains a faithful abelian idempotent. A von Neumann regular right self-injective ring R is said to be of Type II provided R contains a faithful directly finite idempotent but R contains no nonzero abelian idempotents. A von Neumann regular right self-injective ring is said to be of Type III if it contains no nonzero directly finite idempotents.

Definition 2.15 A von Neumann regular right self-injective ring R is said to be of (i) Type I_f if R is of Type I and is directly finite, (ii) Type I_∞ if R is of Type I and is purely infinite, (iii) Type II_f if R is of Type II and is directly finite, (iv) Type II_∞ if R is of Type II and is purely infinite.

Theorem 2.16 *Let R be a von Neumann regular right self-injective ring. Then $R = R_1 \times R_2 \times R_3 \times R_4 \times R_5$, where R_1 is of Type I_f, R_2 is of Type I_∞, R_3 is of Type II_f, R_4 is of Type II_∞, and R_5 is of Type III.*

Definition 2.17 A ring R is called an elementary divisor ring if each square matrix over R admits a diagonal reduction; that is, for each $A \in \mathbb{M}_n(R)$, there exist invertible matrices $P, Q \in \mathbb{M}_n(R)$ such that PAQ is a diagonal matrix.

Lemma 2.18 *If R is an elementary divisor ring, then each element in proper matrix ring $\mathbb{M}_n(R)$ for $n > 1$ is a sum of two unit elements of $\mathbb{M}_n(R)$.*

Proof Let R be an elementary divisor ring and $A \in \mathbb{M}_n(R)$ where $n > 1$. Then there exist diagonal matrices $P, Q \in \mathbb{M}_n(R)$ such that $PAQ = D$,

where D is a diagonal matrix in $\mathbb{M}_n(R)$. Let $D = \begin{pmatrix} a_{11} & \cdots & \cdots & 0 \\ 0 & a_{22} & & \vdots \\ \vdots & & \ddots & \vdots \\ 0 & \cdots & \cdots & a_{nn} \end{pmatrix}$. We

have $D = \begin{pmatrix} a_{11} & 0 & \cdots & 0 & 1 \\ 1 & a_{22} & 0 & \cdots & 0 \\ 0 & 1 & \ddots & \cdots & \vdots \\ \vdots & \vdots & \cdots & a_{n-1n-1} & 0 \\ 0 & 0 & \cdots & 1 & 0 \end{pmatrix} + \begin{pmatrix} 0 & 0 & \cdots & 0 & -1 \\ -1 & 0 & 0 & \cdots & 0 \\ 0 & -1 & \ddots & \cdots & \vdots \\ \vdots & \vdots & \cdots & 0 & 0 \\ 0 & 0 & \cdots & -1 & a_{nn} \end{pmatrix}$.

Set $U = \begin{pmatrix} a_{11} & 0 & \cdots & 0 & 1 \\ 1 & a_{22} & 0 & \cdots & 0 \\ 0 & 1 & \ddots & \cdots & \vdots \\ \vdots & \vdots & \cdots & a_{n-1n-1} & 0 \\ 0 & 0 & \cdots & 1 & 0 \end{pmatrix}$ and $V = \begin{pmatrix} 0 & 0 & \cdots & 0 & -1 \\ -1 & 0 & 0 & \cdots & 0 \\ 0 & -1 & \ddots & \cdots & \vdots \\ \vdots & \vdots & \cdots & 0 & 0 \\ 0 & 0 & \cdots & -1 & a_{nn} \end{pmatrix}$.

It may be checked that U and V are unit elements in $\mathbb{M}_n(R)$. Now, we have $A = P^{-1}UQ^{-1} + P^{-1}VQ^{-1}$. This clearly shows that A is a sum of two unit elements of $\mathbb{M}_n(R)$. □

Lemma 2.19 *[8] Every von Neumann regular right self-injective ring is an elementary divisor ring.*

Theorem 2.20 *Let R be a von Neumann regular right self-injective ring. Then $R = R_1 \times R_2$, where R_1 is an abelian regular self-injective ring and each element of the ring R_2 is a sum of two units.*

Proof By Theorem 2.16, we have that $R \cong S \times T$, where S is purely infinite and T is directly finite. As S is purely infinite, $S_S \cong S_S^2$. Thus, we have $\mathrm{End}(S_S) \cong \mathrm{End}(S_S^{(}2))$. Consequently, we get $S \cong \mathbb{M}_2(S)$. By Lemma 2.19,

S is an elementary divisor ring, and hence, in view of Lemma 2.18, every $A \in \mathbb{M}_2(S)$ is a sum of two units. Hence, we conclude that every element of S is a sum of two units. Now, we consider T. Since T is a directly finite von Neumann regular right self-injective ring, by Theorem 2.16, we have $T \cong T_1 \times T_2$, where T_1 is Type I$_f$ and T_2 is Type II$_f$.

Next, we show that every element of T_2 is sum of two units. Since T_2 is Type II$_f$, it has no nonzero abelian idempotents. Therefore, by [40, proposition 10.28], there exists an idempotent $e_2 \in T_2$ such that $(T_2)_{T_2} \cong 2(e_2 T_2)$ and so $T_2 \cong M_2(e_2 T_2 e_2)$. Again, as T_2 is an elementary divisor ring, by Lemma 2.18, we conclude that every element of $M_2(e_2 T_2 e_2)$, and hence every element of T_2, is a sum of two units.

Finally, we consider T_1. By [40, theorem 10.24], $T_1 \cong \Pi M_n(S_i)$, where each S_i is an abelian regular self-injective ring. If $n > 1$, then clearly each element of $M_n(S_i)$ is a sum of two units by Lemma 2.18. Thus, we have $R = R_1 \times R_2$, where R_1 is an abelian regular self-injective ring and each element in R_2 is a sum of two units. $\square$

Theorem 2.21 *For a right self-injective ring R, the following conditions are equivalent:*

1. *Every element of R is a sum of two units;*
2. *identity element of R is a sum of two units;*
3. *R has no homomorphic image isomorphic to $\mathbb{F}_2$.*

Proof The implications (1) $\Longrightarrow$ (2) $\Longrightarrow$ (3) are obvious.

Now, we proceed to show (3) $\Longrightarrow$ (1).

We know that $R/J(R)$ is a von Neumann regular right self-injective ring. By Theorem 2.20, $R/J(R) = R_1 \times R_2$, where R_1 is an abelian regular self-injective ring and each element in R_2 is a sum of two units. Now, we show that every element of R_1 is a sum of two units. Since by [40, theorem 10.24], $R_1 \cong \Pi M_n(S_i)$, where each S_i is an abelian regular right self-injective ring, it is enough to show that each element of $M_n(S_i)$ is sum of two units. But if $n > 1$, then as S_i are elementary divisor rings, we are through, by Lemma 2.18. So, it is enough to show that every element in an abelian regular ring S_i, which has no homomorphic image isomorphic to $\mathbb{F}_2$, is a sum of two units. Let $a \in S_i$. Suppose, to the contrary, that a is not a sum of two units. Let $\Omega = \{I : I$ is an ideal of S_i and $a + I$ is not a sum of two units in $S_i/I\}$. Clearly, Ω is nonempty, and it can be easily checked that Ω is an inductive set. So, by the Zorn lemma, Ω has a maximal element, say I. Clearly, S_i/I is an indecomposable ring, and hence, has no central idempotent. But, as S_i/I is abelian regular, S_i/I must be a division ring. Since $a + I$ is not a sum of two units in S_i/I, it follows that

$S_i/I \cong \mathbb{F}_2$, a contradiction. Thus, each element of R_1 is a sum of two units. This shows that each element of $R/J(R)$ is a sum of two units. Now, as every element of R is a sum of two units if and only if every element of $R/J(R)$ is a sum of two units, we conclude that each element of R is a sum of two units. This completes the proof. □

Lemma 2.22 *[130, proposition 19] Let M_R be a nonsingular injective module. Then $End(M_R)$ is a Boolean ring if and only if the identity endomorphism of no direct summand of M is a sum of two units.*

Lemma 2.23 *Let T be an abelian regular right self-injective ring. Then $T = T_1 \times T_2$, where each element of T_1 is the sum of two units and T_2 is a Boolean ring (or possibly a zero ring).*

Proof In view of Theorem 2.21, it is enough to show that T has a ring decomposition $T = T_1 \times T_2$, where T_2 is a Boolean ring and the identity in T_1 is a sum of two units in T_1. We shall prove this using a standard Zorn lemma argument. Let $\mathcal{T}$ be the set of all pairs of the form (A, u) where A is a submodule of T_T and u is an automorphism of A such that $I_A - u$ is also an automorphism of A, where I_A is the identity automorphism of A. Clearly, $(0, 0) \in \mathcal{T}$. Then $\mathcal{T}$ has an obvious partial order, i.e. $(A, u) \leq (A', u')$ if $A \subseteq A'$, and u' agrees with u on A. Also, $\mathcal{T}$ is easily seen to be inductive, and so, by Zorn's Lemma, $\mathcal{T}$ has a maximal element, (T_1, v) say. If T_1 is not injective, then v extends to an automorphism v' of an injective hull $E(T_1)$ of T_1 in T_T, and so $(E(T_1), v') \in \mathcal{T}$, violating the maximality of (T_1, v). Thus, T_1 is injective, and so $T = T_1 \oplus T_2$ for some submodule T_2 of T_T. As (T_1, v) is a maximal element of $\mathcal{T}$, it is clear that the identity endomorphism of no direct summand of T_2 is a sum of two units. So, by Lemma 2.22, $End_T(T_2)$ is a Boolean ring. As every idempotent in T is central, the module decomposition $T_T = T_1 \oplus T_2$ gives us a ring decomposition $T = T_1 \times T_2$ with $End_T(T_2) \cong T_2$ a Boolean ring. Also, the identity in T_1 is a sum of two units in T_1. □

In view of the preceding, we have the following.

Theorem 2.24 *Let R be a von Neumann regular right self-injective ring. Then $R = R_1 \times R_2$, where R_1 is a Boolean ring and each element of the ring R_2 is a sum of two units.*

In general, there is no characterization known for when a subring of a von Neumann regular ring is also von Neumann regular. In the theorem that follows, we provide a useful result describing conditions that ensure the subring of a von Neumann regular right self-injective ring is again von Neumann regular.

Theorem 2.25 *Let S be a von Neumann regular right self-injective ring and R be a subring of S that is stable under left multiplication by units of S. Then R is a von Neumann regular ring and $R = R_1 \times R_2$, where R_1 is a Boolean ring and R_2 is a von Neumann regular right self-injective ring.*

Proof By the preceding theorem, we know that $S = S_1 \times S_2$, where S_1 is a Boolean ring and each element of the ring S_2 is a sum of two units. As R is a subring of S, we may view any element a of R as $a = a_1 \times a_2$, where $a_1 \in S_1$ and $a_2 \in S_2$. Since any element $s_2 \in S_2$ is the sum of two units, say $s_2 = t_2 + t_2'$, we may write the element $0 \times s_2 = 1_{S_1} \times t_2 + (-1_{S_2}) \times t_2'$ as the sum of two units of S, and as R is stable under left multiplication by units in S, this means that $0 \times s_2 \in R$ for any $s_2 \in S_2$. Call $R_2 = S_2$ and define

$$R_1 = \{s_1 \in S_1 \mid \exists s_2 \in S_2 \text{ such that } s_1 \times s_2 \in R\}.$$

Then any $s_1 \times 0$ with $s_1 \in R_1$ can be written as $s_1 \times 0 = s_1 \times s_2 - 0 \times s_2$ with $s_1 \times s_2 \in R$. Therefore, $s_1 \times 0 \in R$ for any $s_1 \in R_1$, and we deduce that $R = R_1 \times R_2$. This gives a decomposition for $R = R_1 \times R_2$, with $R_2 = S_2$, and therefore, R_2 is a von Neumann regular right self-injective ring. As every element of R_1 is in S_1, it follows that each element of R_1 is an idempotent, and hence, R_1 is a Boolean ring. Clearly then, R is a von Neumann regular ring. □

Proposition 2.26 *Let S be a right self-injective von Neumann regular ring and R be a subring of S. Assume that R is stable under left multiplication by units of S. If R has no homomorphic images isomorphic to $\mathbb{Z}_2$, then $R = S$.*

Proof Note that our hypothesis implies that S has no homomorphic image isomorphic to $\mathbb{Z}_2$ since, if otherwise, if $\psi \colon S \to \mathbb{Z}_2$ is a ring homomorphism, then $\psi|_R \colon R \to \mathbb{Z}_2$ would give a ring homomorphism, contradicting our assumption. Therefore, each element in S is the sum of two units by Theorem 2.21, and thus, R is invariant under left multiplication by elements of S. But then, calling 1_R to the identity in R, we get that $s = s \cdot 1_R \in R$ for each $s \in S$. Therefore, $R = S$. □

Let S be any ring, and let us denote by $\text{Aut}(S)$, the group of units of S. The canonical ring homomorphism $v \colon \mathbb{Z} \to S$ that takes $1_{\mathbb{Z}}$ to 1_S has kernel 0 or $\mathbb{Z}_n$, for some $n \in \mathbb{N}$. In the first case, S is called a ring of characteristic 0 and, in the other, a ring of characteristic n. Let us denote $\mathbb{Z}$ by $\mathbb{Z}_0$.

Let us denote by S', the image of the group ring $\mathbb{Z}_n[\text{Aut}(S)]$ inside S under the ring homomorphism sending an element of $\text{Aut}(S)$ to the corresponding element in S.

Then S' is a subring of S consisting of those elements that can be written as a finite sum of units of S, where n is the characteristic of the ring S. By construction, the subring S' is invariant under left (or right) multiplication by units of S. The problem stated by Fuchs of characterizing endomorphism rings that are additively generated by automorphisms reduces then to characterizing when $S = S'$. From Theorem 2.25 and Proposition 2.26, we deduce the following partial answers of this question.

Corollary 2.27 *Let S be a von Neumann regular and right self-injective ring of characteristic n. Then S' is also a von Neumann regular ring.*

Corollary 2.28 *Let S be a von Neumann regular and right self-injective ring of characteristic n. If S has no homomorphic images isomorphic to $\mathbb{Z}_2$, then $S = S'$. In particular, this is the case when $n > 0$ and 2 does not divide n.*

Proof The first part is an immediate consequence of Proposition 2.26. For the second part, just note that if $\mathbb{Z}_n$ is the kernel of $\nu \colon \mathbb{Z} \to S$ for some $n > 0$ and 2 does not divide n, then S cannot have any ring homomorphism $\delta \colon S \to \mathbb{Z}_2$ since, if otherwise, 2 and n would belong to $\mathrm{Ker}(\delta \circ \nu)$. And, as 2 and n are coprime, we would deduce that $1 \in \mathrm{Ker}(\delta \circ \nu)$, contradicting that ν is a ring homomorphism. $\square$

As every right self-injective ring is an exchange right quasi-continuous ring, the following result generalizes Theorem 2.21.

Proposition 2.29 *Let M_S be a quasi-continuous module with finite exchange property and $R = \mathrm{End}_S(M)$. If no factor ring of R is isomorphic to $\mathbb{F}_2$, then every element of R is sum of two units.*

Proof Let $\Delta = \{ f \in R \colon \ker f \subseteq_e M \}$. Then Δ is an ideal of R. We have $\overline{R} = R/\Delta \cong R_1 \times R_2$, where R_1 is a von Neumann regular right self-injective ring and R_2 is an exchange ring with no nonzero nilpotent element. We have already shown in Theorem 2.21 that each element of R_1 is a sum of two units. Since R_2 has no nonzero nilpotent element, each idempotent in R_2 is central. Now, if any element $a \in R_2$ is not sum of two units, then, as in the proof of Theorem 2.21, we find an ideal I of R_2 such that $x = a + I \in R_2/I$ is not a sum of two units in R_2/I and R_2/I has no central idempotent. This implies that R_2/I is an exchange ring without any nontrivial idempotent, and hence, it must be a local ring. If $S = R_2/I$, then $x + J(S)$ is not a sum of two units in $S/J(S)$, which is a division ring. Therefore, $S/J(S) \cong \mathbb{F}_2$, a contradiction. Hence, every element of R_2 is also a sum of two units. Therefore, every element of $\overline{R}$ is a sum of two units. Next, we observe that $\Delta \subseteq J(R)$. Suppose to the contrary that $\Delta \nsubseteq J(R)$, then Δ contains a nonzero idempotent, say e. But as

$\text{Ker}(e) \subseteq_e M$, $\text{Ker}(e) = M$ and so $e = 0$, a contradiction. Thus, $\Delta \subseteq J(R)$. This shows that every element of R is a sum of two units. $\square$

Since every continuous module is quasi-continuous and also has the exchange property, we have, in particular, the following.

Theorem 2.30 *Let M be a continuous module. Then each element of* $\text{End}(M)$ *is a sum of two units if* $\text{End}(M)$ *has no homomorphic image isomorphic to* $\mathbb{F}_2$.

In the view of Theorem 1.132, as a particular case of Theorem 2.21, we have the following.

Corollary 2.31 *Each element in the endomorphism ring of a flat cotorsion (in particular, pure-injective) module M is a sum of two units if* $\text{End}(M)$ *has no homomorphic image isomorphic to* $\mathbb{F}_2$.

2.4 Applications of Additive Unit Structure of von Neumann Regular Rings

Using the additive unit structure of von Neumann regular right self-injective rings, we have the following condition under which $\mathcal{X}$-automorphism-invariant modules are $\mathcal{X}$-endomorphism-invariant.

Theorem 2.32 *Let M be $\mathcal{X}$-automorphism-invariant with a monomorphic $\mathcal{X}$-envelope $u\colon M \to X$ such that $\text{End}(X)/J(\text{End}(X))$ is von Neumann regular right self-injective and idempotents lift modulo $J(\text{End}(X))$. If $\text{End}(M)$ has no homomorphic images isomorphic to* $\mathbb{F}_2$, *then M is $\mathcal{X}$-endomorphism-invariant, and therefore,*

$$\text{End}(M)/J(\text{End}(M)) = \text{End}(X)/J(\text{End}(X)).$$

In particular, $\text{End}(M)/J(\text{End}(M))$ is von Neumann regular, right self-injective and idempotents lift modulo $J(\text{End}(M))$.

This is the case when $char(\text{End}(M)) = n > 0$ and $2 \nmid n$.

Proof Assume that $\text{End}(M)$ has no homomorphic images isomorphic to $\mathbb{F}_2$. Then neither has $\text{End}(M)/J(\text{End}(M))$, and thus, we deduce that

$$\text{End}(M)/J(\text{End}(M)) = \text{End}(X)/J(\text{End}(X)),$$

by Proposition 2.26. The proof of Proposition 2.26 shows that $\text{End}(X)/J(\text{End}(X))$ has no homomorphic images isomorphic to $\mathbb{F}_2$, and consequently, it follows that each element in $\text{End}(X)/J(\text{End}(X))$ is the sum of two units. Now, applying Lemma 2.14, we deduce that M is $\mathcal{X}$-endomorphism-invariant.

Finally, if $char(\mathrm{End}(M)) = n > 0$ and $2 \nmid n$, then $\mathrm{End}(M)$ cannot have homomorphic images isomorphic to $\mathbb{F}_2$ for the same reason as in Corollary 2.28.   □

When R is a commutative ring, the preceding theorem has the following consequence.

Corollary 2.33 *If R is a commutative ring with no homomorphic image isomorphic to $\mathbb{F}_2$ and M is an $\mathcal{X}$-automorphism-invariant R-module with a monomorphic $\mathcal{X}$-envelope $u : M \to X$ such that $\mathrm{End}(X)/J(\mathrm{End}(X))$ is von Neumann regular right self-injective and idempotents lift modulo $J(\mathrm{End}(X))$, then M is $\mathcal{X}$-endomorphism-invariant.*

Proof If R is a commutative ring, then there exists a ring homomorphism $f : R \to \mathrm{End}_R(M)$. Now, if there exists a ring homomorphism $g : \mathrm{End}(M) \to \mathbb{F}_2$, then the composition $g \circ f : R \to \mathbb{F}_2$ gives a ring homomorphism, thus contradicting the assumption that R has no homomorphic image isomorphic to $\mathbb{F}_2$. This means that $\mathrm{End}(M)$ has no homomorphic image isomorphic to $\mathbb{F}_2$, and hence, M is $\mathcal{X}$-endomorphism-invariant by the preceding theorem.   □

We are now in a position to construct more examples of $\mathcal{X}$-automorphism-invariant modules that are not $\mathcal{X}$-endomorphism-invariant.

Example 2.34 Let R be the ring of all eventually constant sequences of elements from $\mathbb{F}_2$, the field of two elements. Then R_R is automorphism-invariant but not quasi-injective. Let S be a ring with a right S-module N' which is pure-injective but not injective. Furthermore, assume that S has no homomorphic image isomorphic to $\mathbb{F}_2$. Let $N = E(S) \oplus N'$, $T = R \times S$, $A = R_R$ and $M_T = A \times N$. Note that M_T is invariant under automorphisms of $PE(M)$. It may be observed that M_T is not invariant under endomorphisms of $PE(M)$ because if M is pure-quasi-injective, then A would be quasi-injective, a contradiction. Also, M is not automorphism-invariant because if M was automorphism-invariant, then it would force N to be automorphism-invariant. But as S has no homormorphic image isomorphic to $\mathbb{F}_2$, then N would be quasi-injective and hence injective, as $S \subseteq N$, a contradiction to our assumption that N' is not injective.

We proceed now to characterize when an indecomposable $\mathcal{X}$-automorphism-invariant module is $\mathcal{X}$-endomorphism-invariant.

Lemma 2.35 *Let M be an $\mathcal{X}$-automorphism-invariant module with a monomorphic $\mathcal{X}$-envelope $u : M \to X$ such that $\mathrm{End}(X)/J(\mathrm{End}(X))$ is von Neumann regular right self-injective and idempotents lift modulo $J(\mathrm{End}(X))$. If X is indecomposable, then M is $\mathcal{X}$-endomorphism-invariant.*

Proof Assume X is indecomposable. Then $\text{End}(X)/J(\text{End}(X))$ is a von Neumann regular ring with no nontrivial idempotents, and consequently, $\text{End}(X)/J(\text{End}(X))$ is a division ring. Thus, each element of $\text{End}(X)/J(\text{End}(X))$, and hence of $\text{End}(X)$, is a sum of two or three units [74]. Therefore, M is invariant under any endomorphism of X. □

Our next proposition shows that the converse of the preceding lemma holds under the additional assumption that M is an indecomposable module.

Proposition 2.36 *Let M be an indecomposable $\mathcal{X}$-automorphism-invariant module with a monomorphic $\mathcal{X}$-envelope $u : M \to X$ such that $\text{End}(X)/J(\text{End}(X))$ is von Neumann regular right self-injective and idempotents lift modulo $J(\text{End}(X))$. Then the following statements are equivalent:*

1. *M is $\mathcal{X}$-endomorphism-invariant.*
2. *X is indecomposable.*

Proof Assume M is $\mathcal{X}$-endomorphism-invariant. Then, in particular, M is invariant under any idempotent endomorphism of X. As M is indecomposable, this means that the only idempotents in $\text{End}(X)$ are 0 and 1. So X is indecomposable. The reverse implication follows from the preceding lemma. □

Notes

This results of this chapter are taken from [49], [73], [74] and [53]. Sections 2.1 and 2.2 are completely based on [49]. Section 2.3 is based on [73], [74] and [53], whereas the last section is based on [53].

3

Structure and Properties of Modules Invariant under Automorphisms

Throughout this chapter, $\mathcal{X}$ will denote a class of modules closed under isomorphisms, and direct summands and M will be a module with $u \colon M \to X$, a monomorphic $\mathcal{X}$-envelope such that $\mathrm{End}(X)/J(\mathrm{End}(X))$ is a von Neumann regular right self-injective ring and idempotents lift modulo $J(\mathrm{End}(X))$.

If $f \colon M \to M$ is an endomorphism, then we know by the definition of preenvelope that $u \circ f$ extends to an endomorphism $g \colon X \to X$ such that $g \circ u = u \circ f$. The following easy lemma asserts that this extension is unique modulo the Jacobson radical of $\mathrm{End}(X)$.

Lemma 3.1 *Let $f \in \mathrm{End}(M)$ be any endomorphism, and assume that $g, g' \in \mathrm{End}(X)$ satisfy that $g \circ u = u \circ f = g' \circ u$. Then $g - g' \in J(\mathrm{End}(X))$.*

Proof In order to prove that $g - g' \in J(\mathrm{End}(X))$, we must show that $1 - t \circ (g - g')$ is an automorphism for any $t \in \mathrm{End}(X)$. Let us note that $t \circ (g - g') \circ u = t \circ (g \circ u - g' \circ u) = t \circ u \circ (f - f) = 0$. Therefore, $[1 - t \circ (g - g')] \circ u = u$. Thus, $1 - t \circ (g - g')$ is an automorphism by the definition of envelope. □

The preceding lemma shows that we can define a ring homomorphism

$$\varphi \colon \mathrm{End}(M) \longrightarrow \mathrm{End}(X)/J(\mathrm{End}(X))$$

by the rule $\varphi(f) = g + J(\mathrm{End}(X))$. Call $K = \mathrm{Ker}\,\varphi$. Then φ induces an injective ring homomorphism

$$\Psi \colon \mathrm{End}(M)/K \longrightarrow \mathrm{End}(X)/J(\mathrm{End}(X)),$$

which allows us to identify $\mathrm{End}(M)/K$ with the subring $\mathrm{Im}\,\Psi \subseteq End(X)/J(\mathrm{End}(X))$.

Lemma 3.2 *Assume $j \in J = J(\mathrm{End}(X))$. Then there exists an element $k \in K$ such that $u \circ k = j \circ u$.*

Proof Let $j \in J$. Then $1 - j$ is invertible. Since M is $\mathcal{X}$-automorphism-invariant, there exists a homomorphism $f : M \to M$ such that $u \circ f = (1 - j) \circ u$. Now, as $j = 1 - (1 - j)$, we have that $j \circ u = u - (1 - j) \circ u = u \circ (1 - f)$. Therefore, $u \circ (1 - f) = j \circ u$, and this means that $\varphi(1 - f) = j + J = 0 + J$ and $1 - f \in \operatorname{Ker} \varphi = K$. $\qquad\qquad\square$

3.1 Structure of Modules Invariant under Automorphisms

In this section, we will provide a decomposition theorem for modules invariant under automorphisms of their envelopes and structure of the endomorphism ring of such modules.

Theorem 3.3 *If M is $\mathcal{X}$-automorphism-invariant, then $\operatorname{End}(M)/K$ is von Neumann regular, $K = J(\operatorname{End}(M))$ and idempotents in $\operatorname{End}(M)/J(\operatorname{End}(M))$ lift to idempotents in $\operatorname{End}(M)$.*

Proof Call $S = \operatorname{End}(X)$, $J = J(\operatorname{End}(X))$ and $T = \operatorname{Im} \Psi \cong \operatorname{End}(M)/K$. We want to show that we are in the situation of Theorem 2.25 to get that $\operatorname{End}(M)/K$ is von Neumann regular. In order to prove it, we only need to show that T is invariant under left multiplication by units of S/J. Let $g + J$ be a unit of S/J. Then $g : X \to X$ is an automorphism. And this means that there exists an $f : M \to M$ such that $u \circ f = g \circ u$. Therefore, $\Psi(f + K) = g + J$. Finally, if $\Psi(f' + K)$ is any element of T, we have that

$$(g + J)\Psi(f' + K) = \Psi(f + K)\Psi(f' + K) = \Psi(ff' + K) \in \operatorname{Im} \Psi.$$

This shows that T is invariant under left multiplication by units of S/J; namely, $\operatorname{End}(M)/K$ is von Neumann regular.

As $\operatorname{End}(M)/K$ is von Neumann regular, $J(\operatorname{End}(M)/K) = 0$, and so $J(\operatorname{End}(M)) \subseteq K$. Let us prove the converse. As K is a two-sided ideal of $\operatorname{End}(M)$, it is enough to show that $1 - f$ is invertible for every $f \in K$. Let $f \in K$, and let $g : X \to X$ such that $g \circ u = u \circ f$. As $f \in K$, we get that $g + J = \Psi(f+K) = 0$. Therefore, $1-g$ is a unit in S. As $(1-g)^{-1} : X \to X$ is an automorphism, there exists an $h : M \to M$ such that $(1-g)^{-1} \circ u = u \circ h$. But then

$$u = (1 - g)^{-1} \circ (1 - g) \circ u = (1 - g)^{-1} \circ u \circ (1 - f) = u \circ h \circ (1 - f)$$

and

$$u = (1 - g) \circ (1 - g)^{-1} \circ u = (1 - g) \circ u \circ h = u \circ (1 - f) \circ h.$$

And, as $u: M \to X$ is a monomorphic envelope, this implies that both $(1 - f) \circ h$ and $h \circ (1 - f)$ are automorphisms. Therefore, $1 - f$ is invertible.

Finally, let us prove that idempotents lift modulo $J(\mathrm{End}(M))$. Let us choose an $f \in \mathrm{End}(M)$ such that $f + K = f^2 + K$. Then there exists a homomorphism $g: X \to X$ such that $g \circ u = u \circ f$. Therefore, $g + J = \Psi(f + K)$. And, as $f + K$ is idempotent in $\mathrm{End}(M)/K$, so is $g + J \in S/J$. As idempotents lift modulo J, there exists an $e = e^2 \in S$ such that $g + J = e + J$. Now $g - e \in J$ implies that there exists a $k \in K$ such that $(g - e) \circ u = u \circ k$ by Lemma 3.2. Note that $u \circ (f - k) = e \circ u$ and, thus, $\varphi(f - k) = e + J$.

Therefore,

$$ u \circ (f - k)^2 = e \circ u \circ (f - k) = e^2 \circ u = e \circ u = u \circ (f - k). $$

And, as u is monic, we get that $(f - k)^2 = f - k$. This shows that idempotents lift modulo $J(\mathrm{End}(M))$. □

Lemma 3.4 *A direct summand of an $\mathcal{X}$-automorphism-invariant module is also $\mathcal{X}$-automorphism-invariant.*

Proof Let us write $M = N \oplus L$, and let $u_N: N \to X_N$ and $u_L: L \to X_L$ be the $\mathcal{X}$-envelopes of N, L, respectively. Then the induced morphism $u: M = N \oplus L \to X_N \oplus X_L$ is also an $\mathcal{X}$-envelope by [137, theorem 1.2.5]. Let now $f: X_N \to X_N$ be any automorphism and consider the diagonal automorphism $(f, 1_{X_L}): X_N \oplus X_L \to X_N \oplus X_L$. As M is $\mathcal{X}$-automorphism-invariant, we get that $(f, 1_{X_L})(M) \subseteq M$. But this means that $f(N) \subseteq N$ by construction. □

Lemma 3.5 *A direct summand of an $\mathcal{X}$-endomorphism-invariant module is also $\mathcal{X}$-endomorphism-invariant.*

Proof Let M be an $\mathcal{X}$-endomorphism-invariant module and N be a direct summand of M. So there exists a module K such that $M = N \oplus K$. Thus, $X(M) = X(N) \oplus X(K)$. Let $f: X(N) \to X(N)$ be an endomorphism of $X(N)$. So $\iota_{X(N)} \circ f \circ \pi_{X(N)}$ is an endomorphism of $X(M)$, where $\iota_{X(N)}: X(N) \to X(M)$ is the inclusion and $\pi_{X(N)}: X(M) \to X(N)$ is the canonical projection. We clearly have $v_N \circ \pi_N = \pi_{X(N)} \circ v_M$ and $v_M \circ \iota_N = \iota_{X(N)} \circ v_N$ with $\iota_N: N \to M$ being the inclusion and $\pi_N: M \to N$ being the canonical projection. As M is $\mathcal{X}$-endomorphism-invariant, there exists $h: M \to M$ such that $v_M \circ h = \iota_{X(N)} \circ f \circ \pi_{X(N)} \circ v_M$. We deduce that $g = \pi_N \circ h \circ \iota_N: N \to N$ is an endomorphism of N such that $v_N \circ g = f \circ v_N$. So N is an $\mathcal{X}$-endomorphism-invariant module. □

Let $M = N \oplus L$ be a decomposition of a module M into two direct summands and call $v_N: N \to M, v_L: L \to M, \pi_N: M \to N, \pi_L: M \to L$ the associated structural injections and projections. We may associate to any homomorphism $f \in \mathrm{Hom}(N, L)$, the endomorphism $v_L \circ f \circ \pi_N$ of M and thus identify $\mathrm{Hom}(N, L)$ with a subset of $\mathrm{End}(M)$. Similarly, we identify $\mathrm{Hom}(L, N)$ with a subset of $\mathrm{End}(M)$ as well. We will use these identifications in the following theorem.

Theorem 3.6 *If M is X-automorphism-invariant and every direct summand of M has an X-envelope, then M admits a decomposition $M = N \oplus L$ such that*

1. *N is a semi-Boolean module.*
2. *L is X-endomorphism-invariant and $\mathrm{End}(L)/J(\mathrm{End}(L))$ is von Neumann regular, right self-injective and idempotents lift modulo $J(\mathrm{End}(L))$.*
3. *Both $\mathrm{Hom}_R(N, L)$ and $\mathrm{Hom}_R(L, N)$ are contained in $J(\mathrm{End}(M))$.*

In particular, $\mathrm{End}(M)/J(\mathrm{End}(M))$ is the direct product of a Boolean ring and a right self-injective von Neumann regular ring.

Proof Let us call $S = \mathrm{End}(X)$ and decompose, as in Lemma 2.16, $S/J(S) = T_1 \times T_2$, where T_1 is a Boolean ring and T_2 is a right self-injective ring in which every element is the sum of two units. Let $\Psi : \mathrm{End}(M)/J(\mathrm{End}(M)) \rightarrow S/J(S)$ be the injective ring homomorphism defined in the beginning of this section. Identifying $\mathrm{End}(M)/J(\mathrm{End}(M))$ with $\mathrm{Im}(\Psi)$, we get that $\mathrm{End}(M)/J(\mathrm{End}(M))$ is a subring of $S/J(S)$ invariant under left multiplication by units of $S/J(S)$. Now, using Theorem 2.25, we obtain that $\mathrm{End}(M)/J(\mathrm{End}(M)) = R_1 \times R_2$, where R_1 is a Boolean ring and R_2 is a right self-injective von Neumann regular ring that is invariant under left multiplication by elements in $S/J(S)$. Let $e + J(\mathrm{End}(M))$ be a central idempotent of $\mathrm{End}(M)/J(\mathrm{End}(M))$ such that $R_1 = e \cdot \mathrm{End}(M)/J(\mathrm{End}(M))$ and $R_2 = (1 - e) \cdot \mathrm{End}(M)/J(\mathrm{End}(M))$. As idempotents lift modulo $J(\mathrm{End}(M))$, we may choose e to be an idempotent of $\mathrm{End}(M)$. Then call $N = eM$ and $L = (1 - e)M$. Note first that both $\mathrm{Hom}_R(N, L)$ and $\mathrm{Hom}_R(L, N)$ are contained in $J(\mathrm{End}(M))$ since

$$\mathrm{End}(M)/J(\mathrm{End}(M)) = [e \cdot \mathrm{End}(M)/J(\mathrm{End}(M))]$$
$$\times [(1 - e) \cdot \mathrm{End}(M)/J(\mathrm{End}(M))].$$

This shows (2). Moreover, as $e + J(\mathrm{End}(M))$ is central in $\mathrm{End}(M)/J(\mathrm{End}(M))$, we get that

$$\mathrm{End}(N)/J(\mathrm{End}(N)) \cong R_1$$

and

$$\mathrm{End}(L)/J(\mathrm{End}(L)) \cong R_2.$$

Now, as both N, L are direct summands of M, they are $\mathcal{X}$-automorphism-invariant by Lemma 3.4. In particular, we get that idempotents in $\text{End}(N)/J(\text{End}(N))$ and in $\text{End}(L)/J(\text{End}(L))$ lift to idempotents in $\text{End}(N)$ and $\text{End}(L)$, respectively. As each element of $\text{End}(N)/J(\text{End}(N))$ is an idempotent and idempotents lift modulo $J(\text{End}(N))$, it turns out that each element of $\text{End}(N)$ is an idempotent as well. This shows that N is semi-Boolean. Finally, as R_2 is invariant under left multiplication by elements in $S/J(S)$, it follows that L is $\mathcal{X}$-endomorphism-invariant. This proves (2), thus completing the proof of the theorem. □

As a consequence, we have the following characterization of indecomposable $\mathcal{X}$-automorphism-invariant modules that are not $\mathcal{X}$-endomorphism invariant.

Theorem 3.7 *Let M be an indecomposable $\mathcal{X}$-automorphism-invariant module with a monomorphic $\mathcal{X}$-envelope $u: M \to X$ such that $\text{End}(X)/J(\text{End}(X))$ is von Neumann regular right self-injective and idempotents lift modulo $J(\text{End}(X))$. Assume that M is not $\mathcal{X}$-endomorphism-invariant. Then $\text{End}(M)/J(\text{End}(M)) \cong \mathbb{F}_2$ and $\text{End}(X)/J(\text{End}(X))$ has a homomorphic image isomorphic to $\mathbb{F}_2 \times \mathbb{F}_2$.*

Proof By Theorem 3.6, $\text{End}(M)/J(\text{End}(M))$ is a von Neumann regular ring. As M is indecomposable, it follows that $\text{End}(M)/J(\text{End}(M))$ is a von Neumann regular ring with no nontrivial idempotents, and consequently, $\text{End}(M)/J(\text{End}(M))$ is a division ring. Now, by Theorem 2.32, we know that $\text{End}(M)$ has a homomorphic image isomorphic to $\mathbb{F}_2$. Thus, we have $\text{End}(M)/J(\text{End}(M)) \cong \mathbb{F}_2$. By Theorem 2.30, it is known that if $\text{End}(X)/J(\text{End}(X))$ has no homomorphic image isomorphic to $\mathbb{F}_2 \times \mathbb{F}_2$, then each element of $\text{End}(X)$ is a sum of two or three units. This would make M an $\mathcal{X}$-endomorphism-invariant module, a contradiction to our assumption. This shows that $\text{End}(X)/J(\text{End}(X))$ has a homomorphic image isomorphic to $\mathbb{F}_2 \times \mathbb{F}_2$. □

3.2 Properties of Modules Invariant under Automorphisms

Theorem 3.8 *If M is $\mathcal{X}$-automorphism-invariant, then M satisfies the finite exchange property.*

Proof We have shown in Theorem 3.3 that $\text{End}(M)/J(\text{End}(M))$ is a von Neumann regular ring and idempotents lift modulo $J(\text{End}(M))$. Then $\text{End}(M)$

is an exchange ring by [84, proposition 1.6]. This proves that M has the finite exchange property. $\qquad\square$

Theorem 3.9 *Let M be an $\mathcal{X}$-automorphism-invariant module, and assume every direct summand of M has an $\mathcal{X}$-envelope. Assume further that for $\mathcal{X}$-endomorphism-invariant modules, the finite exchange property implies the full exchange property. Then M satisfies the full exchange property.*

Proof By the preceding theorem, we have the decomposition $M = N \oplus L$, where N is a semi-Boolean module and L is an $\mathcal{X}$-endomorphism-invariant module. Now, in Corollary 3.8, we have seen that M satisfies the finite exchange property. Therefore, both N and L satisfy the finite exchange property. By our hypothesis, L satisfies the full exchange property. It is known that for a square-free module, the finite exchange property implies the full exchange property. Since a direct sum of two modules with the full exchange property also has the full exchange property, it follows that M satisfies the full exchange property. $\qquad\square$

Theorem 3.10 *Let M be an $\mathcal{X}$-automorphism-invariant module, and assume that every direct summand of M has an $\mathcal{X}$-envelope. Then M is a clean module.*

Proof By Theorem 3.6, $\mathrm{End}(M)/J(\mathrm{End}(M))$ is the direct product of a Boolean ring and a right self-injective von Neumann regular ring. We know that Boolean rings and right self-injective rings are clean. Since the direct product of clean rings is clean, it follows that $\mathrm{End}(M)/J(\mathrm{End}(M))$ is clean. We have also shown in Theorem 3.3 that idempotents in $\mathrm{End}(M)/J(\mathrm{End}(M))$ lift to idempotents in $\mathrm{End}(M)$. Therefore, $\mathrm{End}(M)$ is a clean ring [84, p. 272]. Thus, M is a clean module. $\qquad\square$

Lemma 3.11 *Let M be an $\mathcal{X}$-automorphism-invariant module with $u: M \longrightarrow X(M)$, a monomorphic $\mathcal{X}$-envelope such that $\mathrm{End}(X(M))/J(\mathrm{End}(X(M)))$ is a von Neumann regular right self-injective ring and idempotents lift modulo $J(\mathrm{End}(X(M)))$. Then M is directly infinite if and only if $X(M)$ is directly infinite.*

Proof Assume M is directly infinite. Let $S = \mathrm{End}(X(M))$, and let $s, t \in S$ such that $t \circ s = 1_{X(M)}$. Then $s: X(M) \longrightarrow X(M)$ is a monomorphism. In the view of Theorem 3.18, we know that M is $\mathcal{X}$-monomorphism-invariant. So there exists $f \in \mathrm{End}(M) = T$ such that $u \circ f = s \circ u$. Thus, we have $\Psi(f + K) = s + J(S)$. Therefore, f is a monomorphism. As $K = J(T)$, T/K is von Neumann regular and idempotents lift, there exists an idempotent $e \in T$ such that $Te + K = Tf + K$. This means that there exists $t' \in T$ such that $f = t' \circ e + k$ with $k \in K$, and thus, $f \circ (1 - e) = k \circ (1 - e)$.

Therefore, $\Psi(f + K) \circ \Psi((1 - e) + K) = 0 + J(S)$, and, as $\Psi(f + K)$ is monic in $S/J(S)$, this means that $\Psi((1 - e) + K) = 0 + J(S)$. Since again Ψ is monic, $(1 - e) \in K = J(T)$. Then e is unit in T. Since $M = e(M) \oplus (1 - e)(M)$, $(1 - e)(M) = 0$, and so $e = 1_M$. Therefore, $T = Tf + K$. Now there exists an element $f' \in T$ such that $1_M - f' \circ f \in J(T)$, and thus, $f' \circ f$ is an automorphism. As M is directly finite, this means that f must be an automorphism. Therefore, $s + J(S) = \Psi(f + K)$ is also an automorphism. Then s is an automorphism. As $t \circ s = 1_{X(M)}$, we get that $s \circ t = 1_{X(M)}$, too. This shows that $X(M)$ is directly finite.

Conversely, suppose $X(M)$ is directly finite. We wish to show that M is directly finite. Assume, to the contrary, that M is not directly finite. Then $M \cong M \oplus N$ for some nonzero submodule N. Let $u : M \longrightarrow X(M)$ and $v : N \longrightarrow X(N)$ be $\mathcal{X}$-envelopes of M and N, respectively. Then $u \oplus v : M \oplus N \longrightarrow X(M) \oplus X(N)$ is an $\mathcal{X}$-envelope of $M \oplus N$. Clearly, then $X(M) \cong X(M) \oplus X(N)$. This yields a contradiction to our assumption that $X(M)$ is directly finite. Thus, we have that M is directly finite if and only if $X(M)$ is directly finite.   □

Lemma 3.12 *Let M be an $\mathcal{X}$-automorphism-invariant module with $u : M \longrightarrow X(M)$ a monomorphic $\mathcal{X}$-envelope such that $\mathrm{End}(X(M))/J(\mathrm{End}(X(M)))$ is a von Neumann regular right self-injective ring and idempotents lift modulo $J(\mathrm{End}(X(M)))$. If M is directly finite, then $\mathrm{End}(M)/J(\mathrm{End}(M))$ is unit-regular.*

Proof We know that $\mathrm{End}(M)/J(\mathrm{End}(M)) = R_1 \times R_2$ where R_1 is a Boolean ring and R_2 is a von Neumann regular right self-injective ring. If M is a directly finite module then $(\mathrm{End}(M))$, and hence $\mathrm{End}(M)/J(\mathrm{End}(M))$, is a directly finite ring. Thus, R_2 is a directly finite von Neumann regular right self-injective ring and hence a unit-regular ring. Clearly, R_1 is a unit-regular ring being a Boolean ring. Since a product of two unit-regular rings is unit-regular, $\mathrm{End}(M)/J(\mathrm{End}(M))$ is unit-regular.   □

Theorem 3.13 *Let M be an $\mathcal{X}$-automorphism-invariant module with a monomorphic $\mathcal{X}$-envelope $u : M \longrightarrow X(M)$ such that $\mathrm{End}(X(M))/J(\mathrm{End}(X(N)))$ is a von Neumann regular right self-injective ring and idempotents lift modulo $J(\mathrm{End}(X(M)))$. Then the following are equivalent:*

1. *M is directly finite.*
2. *M has the internal cancellation property.*
3. *M has the cancellation property.*
4. *M has the substitution property.*
5. *$X(M)$ is directly finite.*

6. $X(M)$ *has the internal cancellation property.*
7. $X(M)$ *has the cancellation property.*
8. $X(M)$ *has the substitution property.*

Proof (1) $\Rightarrow$ (2). By Lemma 3.12, $\mathrm{End}(M)/J(\mathrm{End}(M))$ is unit-regular, and therefore, $\mathrm{End}(M)$ is a ring with internal cancellation. Now, by [54], M has the internal cancellation property.

(2) $\Rightarrow$ (3). By Theorem 3.8, M has the finite exchange property. By [138], M has the cancellation property.

(3) $\Rightarrow$ (1), by definitions.

(2) $\Leftrightarrow$ (4), by [138], since M has the finite exchange property.

The equivalence of (5), (6), (7) and (8) with rest of the statements follows from Lemma 3.11. □

3.3 Applications

We are going to show how our results can be applied to a wide variety of classes of modules, obtaining interesting consequences for them.

Theorem 3.14 *Let* $\mathcal{X}$ *be the class of injective modules and* M *be an* $\mathcal{X}$-*automorphism-invariant module. Then*

1. $\mathrm{End}(M)/J(\mathrm{End}(M))$ *is von Neumann regular and idempotents lift modulo* $J(\mathrm{End}(M))$.
2. $M = N \oplus L$ *where* N *is semi-Boolean and* L *is quasi-injective.*
3. M *satisfies the full exchange property.*
4. M *is a clean module.*
5. *If* $\mathrm{End}(M)$ *has no homomorphic images isomorphic to* $\mathbb{F}_2$, *then* M *is quasi-injective.*

Proof Let E be the injective envelope of M. Then $S = \mathrm{End}(E)$ satisfies that $S/J(S)$ is von Neumann regular, right self-injective and idempotents lift modulo $J(S)$ (see, in particular, Theorem 1.118). Therefore:

1. By Theorem 3.3, it follows that $\mathrm{End}(M)/J(\mathrm{End}(M))$ is von Neumann regular and idempotents lift modulo $J(\mathrm{End}(M))$.
2. By Theorem 3.6, $M = N \oplus L$, where N is semi-Boolean and L is quasi-injective.
3. Since every quasi-injective module satisfies the full exchange property (see Theorem 1.98), we deduce from Theorem 3.9 that M also satisfies the full exchange property.

4. By Theorem 3.10, M is a clean module.
5. By Theorem 2.32, M is quasi-injective.   □

Theorem 3.15 *Let* $\mathcal{X}$ *be the class of pure-injective modules and* M *an* $\mathcal{X}$*-automorphism-invariant module. Then:*

1. $\mathrm{End}(M)/J(\mathrm{End}(M))$ *is von Neumann regular and idempotents lift modulo* $J(\mathrm{End}(M))$.
2. $M = N \oplus L$, *where* N *is semi-Boolean and* L *is invariant under endomorphisms of its pure-injective envelope.*
3. M *satisfies the full exchange property.*
4. M *is a clean module.*
5. *If* $\mathrm{End}(M)$ *has no homomorphic images isomorphic to* $\mathbb{F}_2$, *then* M *is invariant under endomorphisms of its pure-injective envelope.*

Proof Let E be the pure-injective envelope of M. Then $S = \mathrm{End}(E)$ satisfies that $S/J(S)$ is von Neumann regular, right self-injective and idempotents lift modulo $J(S)$. Therefore:

1. By Theorem 3.3, it follows that $\mathrm{End}(M)/J(\mathrm{End}(M))$ is von Neumann regular and idempotents lift modulo $J(\mathrm{End}(M))$.
2. By Theorem 3.6, $M = N \oplus L$, where N is semi-Boolean and L is invariant under endomorphisms of its pure-injective envelope. In fact, L turns out to be strongly invariant in its pure-injective envelope in the sense of [61, definition, p. 430].
3. It follows from [61, theorem 11] that a module invariant under endomorphisms of its pure-injective envelope satisfies the full exchange property. Therefore, we deduce from Theorem 3.9 that M also satisfies the full exchange property.
4. By theorem 3.10, M is a clean module.
5. If follows from Theorem 2.32.   □

It is well known that if $(\mathcal{F},\mathcal{C})$ is a cotorsion pair and $u\colon M \to C(M)$ is a $\mathcal{C}$-envelope, then $\mathrm{Coker}(u) \in \mathcal{F}$ (see e.g. [137]). In particular, if $M \in \mathcal{F}$, then $C(M) \in \mathcal{F}$ (since $\mathcal{F}$ is closed under extensions), and thus, $C(M) \in \mathcal{F} \cap \mathcal{C}$. Dually, if $p\colon F(M) \to M$ is an $\mathcal{F}$-cover, then $\mathrm{Ker}\, p \in \mathcal{C}$ and, if $M \in \mathcal{C}$, this means that $F(M) \in \mathcal{C}$, as $\mathcal{C}$ is also closed under extensions. Therefore, we also get that $F(M) \in \mathcal{F} \cap \mathcal{C}$. As a consequence, we can apply our previous results to this situation.

Theorem 3.16 *Let* $(\mathcal{F},\mathcal{C})$ *be a cotorsion pair such that* $\mathcal{F}$ *is closed under direct limits and every module has a* $\mathcal{C}$*-envelope. If* $M \in \mathcal{F}$ *is* $\mathcal{C}$*-automorphism-invariant, then*

1. $\text{End}(M)/J(\text{End}(M))$ *is von Neumann regular and idempotents lift modulo* $J(\text{End}(M))$. *Consequently, M satisfies the finite exchange property.*
2. $M = N \oplus L$, *where N is semi-Boolean, and L is C-endomorphism-invariant.*
3. *M is a clean module.*

In particular, the preceding theorem applies to the case of the Enochs cotorsion pair, in which $\mathcal{F}$ is the class of flat modules and $\mathcal{C}$, the class of cotorsion modules.

3.4 Modules Invariant under Monomorphisms

Definition 3.17 A module M with an $\mathcal{X}$-envelope $u \colon M \to X(M)$ is said to be $\mathcal{X}$-monomorphism-invariant if for any monomorphism $g \colon X \to X$ there exists an endomorphism $f \colon M \to M$ such that $u \circ f = g \circ u$.

Theorem 3.18 *Let M be a module with a monomorphic $\mathcal{X}$-envelope $u \colon M \to X(M)$ such that* $\text{End}(X(M))/J(\text{End}(X(M)))$ *is a von Neumann regular right self-injective ring and idempotents lift modulo* $J(\text{End}(X(M)))$. *Then the following are equivalent:*

1. *M is an $\mathcal{X}$-automorphism-invariant module.*
2. *M is an $\mathcal{X}$-monomorphism-invariant module.*

Proof (1) $\implies$ (2). Let M be $\mathcal{X}$-automorphism-invariant. Call $X = X(M)$. As X is obviously $\mathcal{X}$-automorphism-invariant, we have that $X = X_1 \oplus X_2$, where X_1 is a semi-Boolean module, X_2 is $\mathcal{X}$-endomorphism-invariant, $\text{Hom}(X_1, X_2)$, $\text{Hom}(X_2, X_1) \subseteq J(\text{End}(X))$ and

$$\text{End}(X)/J(\text{End}(X)) = \text{End}(X_1)/J(\text{End}(X_1)) \times \text{End}(X_2)/J(\text{End}(X_2)),$$

where $\text{End}(X_1)/J(\text{End}(X_1))$ is a Boolean ring, and each element in the ring $\text{End}(X_2)/J(\text{End}(X_2))$ is the sum of two units. Call $S = \text{End}(X)$, $S_1 = \text{End}(X_1)$ and $S_2 = \text{End}(X_2)$. Let $f \colon X \to X$ be a monomorphism. As $\bar{S} = \bar{S}_1 \times \bar{S}_2$, there exist $\bar{f}_1 \in \bar{S}_1$ and $\bar{f}_2 \in \bar{S}_2$ such that $\bar{f} = \bar{f}_1 \times \bar{f}_2$. Moreover, each $\bar{f}_i \colon X_i \to X_i$ is a monomorphism. As $\bar{S}_1$ is a Boolean ring, $\bar{f}_1 \in \bar{S}_1$ is a unit. On the other hand, $\bar{f}_2 \in \bar{S}_2$, so we can write $\bar{f}_2$ as the sum of two units, say $\bar{f}_2 = \bar{f}_2' + \bar{f}_2''$. And $\bar{f}_2''$ is again a sum of two units, say $\bar{f}_2'' = \bar{f}_3' + \bar{f}_3''$. Now $\bar{f} = \bar{f}_1 \times \bar{f}_2 = (\bar{1}_{\bar{S}_1} \times \bar{f}_2') + (-\bar{1}_{\bar{S}_1} \times \bar{f}_3') + (\bar{f}_1 \times \bar{f}_3'')$.

Set $\bar{\gamma}_1 = \bar{1}_{\bar{S}_1} \times \bar{f}_2'$, $\bar{\gamma}_2 = -\bar{1}_{\bar{S}_1} \times \bar{f}_3'$, and $\bar{\gamma}_3 = \bar{f}_1 \times \bar{f}_3''$. Then $\bar{f} = \bar{\gamma}_1 + \bar{\gamma}_2 + \bar{\gamma}_3$. Thus, $\bar{f}$ is the sum of three units $\bar{\gamma}_1$, $\bar{\gamma}_2$ and $\bar{\gamma}_3$ in $\bar{S}$. Let us lift the three $\bar{\gamma}_i \in \bar{S}$ to units $\gamma_i \in S$. Then $f = \gamma_1 + \gamma_2 + \gamma_3 + j$ for some

$j \in J(S)$. As M is $\mathcal{X}$-automorphism-invariant, $\gamma_i(M) \subseteq M$ for $i = 1, 2, 3$. Also, as $j \in J(S)$, $1 - j$ is an automorphism, consequently, M is invariant under $1 - j$, and hence under j. Thus, it follows that M is invariant under f.

(2) $\Longrightarrow$ (1). This implication is obvious. $\qquad\qquad\qquad\square$

In particular, we have the following corollary.

Corollary 3.19 *For a module M, we have the following:*

1. *M is automorphism-invariant if and only if M is invariant under monomorphisms of its injective envelope.*
2. *M is pure-automorphism-invariant if and only if M is invariant under monomorphisms of its pure-injective envelope.*
3. *If M is flat, then M is invariant under automorphisms of its cotorsion envelope if and only if it is invariant under all monomorphisms of its cotorsion envelope.*

Notes

This chapter is based on [49] and [50]. The results mentioned in Sections 3.1, 3.2 and 3.3 are taken from [49], whereas the results of Section 3.4 are taken from [50].

4

Automorphism-Invariant Modules

When $\mathcal{X}$ is the class of injective modules, then $\mathcal{X}$-automorphism-invariant modules are simply called automorphism-invariant modules. These modules were first studied by Dickson and Fuller in [22]. Lee and Zhou in [79] called such modules auto-invariant modules. In this chapter, we will see some equivalent characterizations for a module to be invariant under automorphisms of its injective envelope. We will explore conditions when automorphism-invariant modules are precisely the quasi-injective modules.

4.1 Some Characterizations of Automorphism-Invariant Modules

In this section, we will give some equivalent characterizations for a module to be automorphism-invariant.

Theorem 4.1 *For a right R-module M with injective envelope E, the following statements are equivalent.*

1. *M is an automorphism-invariant module.*
2. *M is a characteristic submodule of E.*
3. $M = \alpha(M) = \alpha^{-1}(M)$ *for every automorphism α of E.*
4. *Every isomorphism between any two essential submodules of the module M can be extended to an endomorphism of the module M.*
5. *Every isomorphism between any two essential submodules of M can be extended to an automorphism of the module M.*
6. *M is a characteristic submodule of some injective module G.*

Proof Clearly, (1) and (2) are equivalent. The implications $(3) \Rightarrow (2)$, $(6) \Rightarrow (2)$ and $(5) \Rightarrow (4)$ are obvious.

(4) $\Rightarrow$ (2). Let α be an automorphism of the module E, $X' = M \cap \alpha(M) \subseteq M$, and let $X = \alpha^{-1}(X') \subseteq M$. Since $\alpha(M)$ is an essential submodule of the module $G = \alpha(E)$, we have that X' is an essential submodule of the module M. In addition, $X = \alpha^{-1}(X')$ is an essential submodule of the module $E = \alpha^{-1}Q$. Since α induces the isomorphism from the module X onto X', it follows from (4) that there exists an endomorphism β of the module M that coincides with α at X. We assume that $z \in M \cap (\alpha - \beta)(M)$. Then $z = (\alpha - \beta)(y)$, where $y \in M$. Then $\alpha(y) = \beta(y) - z \in M$. Thus, $y \in X$. However, $(\alpha - \beta)(y) = z = 0$ by the construction. Therefore, $M \cap (\alpha - \beta)(M) = 0$. Thus, $(\alpha - \beta)(M) = 0$, since M is an essential submodule of E. Therefore, $\alpha(M) \subseteq M$.

(2) $\Rightarrow$ (3). Since M is an automorphism-invariant module, $\alpha(M) \subseteq M$ and $\alpha^{-1}(M) \subseteq M$. Therefore, $M = \alpha(M) = \alpha^{-1}(M)$.

(3) $\Rightarrow$ (5). Let X and X' be two essential submodules of the module M, and let $\alpha' : X \to X'$ be an isomorphism. Since E is an injective module, there exists an endomorphism α of the module E that coincides with α' ate X. Since X is an essential submodule in E and $X \cap \text{Ker} \, \alpha = 0$, we have that α is a monomorphism. Therefore, $\alpha(E) \cong E$, and $\alpha(E)$ is an injective module. Then $\alpha(E)$ is a direct summand of the module E. In addition, $\alpha(E)$ contains the submodule $\alpha(X)$, and $X' = \alpha X$. Therefore, $\alpha(E)$ is an essential direct summand of the module E. Then $\alpha(E) = E$, and α is an automorphism of the module E. By (3), the restriction of α to M is an endomorphism of the module M that coincides with α' at X.

(2) $\Rightarrow$ (6). Let $G = E \oplus F$. Every automorphism α of the module E can be extended to an automorphism β of the module G with the use of the relation $\beta(x + y) = \alpha(x) + y$. By the assumption, $\beta(M) \subseteq M$. Then $\alpha(M) = \beta(M) \subseteq M$. $\qquad\qquad\qquad\square$

Definition 4.2 A module M is called pseudo-injective if, given any submodule A of M, any monomorphism $f : A \to M$ can be extended to an endomorphism of M.

Jain and Singh in [63] introduced the notion of pseudo-injective modules. It is quite obvious that quasi-injective modules are pseudo-injective. Teply gave construction for example of pseudo-injective modules that are not quasi-injective [100].

In the next theorem, we will show that automorphism-invariant modules coincide with pseudo-injective modules. This was first proved by Er, Singh and Srivastava in [32], but the proof we give in what follows is based on [50].

Theorem 4.3 *A module M is automorphism-invariant if and only if it is pseudo-injective.*

Proof Let M be an automorphism-invariant module. Let N be a submodule of M, and let $f : N \to M$ be a monomorphism. Call $E = E(M)$. It is well known that $\text{End}(E)/J(\text{End}(E))$ is von Neumann regular, right self-injective and idempotents lift modulo $J(\text{End}(E))$. Since injective modules are obviously automorphism-invariant, from Theorem 3.6, we have that $E = E_1 \oplus E_2$, where $\text{Hom}(E_1, E_2), \text{Hom}(E_2, E_1) \subseteq J(\text{End}(E))$ and

$$\text{End}(E)/J(\text{End}(E)) = \text{End}(E_1)/J(\text{End}(E_1)) \times \text{End}(E_2)/J(\text{End}(E_2))$$

such that $\text{End}(E_1)/J(\text{End}(E_1))$ is a Boolean ring and each element in the ring $\text{End}(E_2)/J(\text{End}(E_2))$ is the sum of two units. Call $S = \text{End}(E)$, $S_1 = \text{End}(E_1)$ and $S_2 = \text{End}(E_2)$. Let $v : E(N) \to E$ be the inclusion, and let $p : E \to E(N)$ be an epimorphism that splits v. Then $e = v \circ p \in S$ is an idempotent such that $E(N) = eE$. By injectivity, f extends to a monomorphism $g : E(N) \to E$. This monomorphism g splits as $E(N)$ is injective. So there exists an epimorphism $\delta : E \to E(N)$ such that $\delta \circ g = 1_{E(N)}$. Call $h = g \circ p \in S$. We claim that $\bar{h}|_{\bar{e}\bar{S}} : \bar{e}\bar{S} \to \bar{S}$ is a monomorphism. Let $x \in S$ such that $\bar{e}\bar{x} \in \text{Ker}(\bar{h})$. This means that $h \circ e \circ x = g \circ p \circ e \circ x$ has essential kernel. And, as g is a monomorphism, we deduce that $p \circ e \circ x = p \circ v \circ p \circ x = p \circ x$ also has essential kernel. So $e \circ x = v \circ p \circ x$ has essential kernel, too. This shows that $\bar{e}\bar{x} = 0$ and, thus, $\bar{h}|_{\bar{e}\bar{S}}$ is monic.

As $\bar{S} = \bar{S}_1 \times \bar{S}_2$, there exist idempotents $\bar{e}_1 \in \bar{S}_1$ and $\bar{e}_2 \in \bar{S}_2$ such that $\bar{e} = \bar{e}_1 \times \bar{e}_2 \in \bar{S}_1 \times \bar{S}_2$ and homomorphisms $\bar{h}_1 : \bar{S}_1 \to \bar{S}_1$ and $\bar{h}_2 : \bar{S}_2 \to \bar{S}_2$ such that $\bar{h} = \bar{h}_1 \times \bar{h}_2$. Moreover, $\bar{h}_i|_{\bar{e}_i\bar{S}_i} : \bar{e}_i\bar{S}_i \to \bar{S}_i$ is a monomorphism and $\bar{h}_i|_{(1-\bar{e}_i)\bar{S}_i} = 0$ for $i = 1, 2$. As $\text{Im}(\bar{h}) \cong \bar{e}\bar{S}$, it is a direct summand of $\bar{S}$. So there exists an idempotent $\bar{e}' \in \bar{S}$ such that $\text{Im}(\bar{h}) = \bar{e}'\bar{S}$. And again, $\bar{e}' = \bar{e}'_1 \times \bar{e}'_2$ for idempotents $\bar{e}'_1 \in \bar{S}_1$ and $\bar{e}'_2 \in \bar{S}_2$. Also, we have $\text{Ker}(\bar{h}_1) = (1 - \bar{e}_1)S$ as $\bar{e}_1 \in \bar{S}_1$ is central because $\bar{S}_1$ is a Boolean ring and $(1 - e)h = 0$. This yields $\bar{e}_1 = \bar{e}'_1$.

Call $\bar{h}'_1 : \bar{S}_1 \to \bar{S}_1$ the homomorphism defined by $\bar{h}'_1|_{\bar{e}_1\bar{S}_1} = \bar{h}_1|_{\bar{e}_1\bar{S}_1}$ and $\bar{h}'_1|_{(1-\bar{e}_1)\bar{S}_1} = 1_{(1-\bar{e}_1)\bar{S}_1}$. By construction, $\bar{h}'_1$ is an automorphism in $\bar{S}_1$. On the other hand, $\bar{h}_2 \in \bar{S}_2$, so we can write $\bar{h}_2$ as the sum of two automorphisms, say $\bar{h}_2 = \bar{h}'_2 + \bar{h}''_2$. And again, $\bar{h}''_2$ can be written as the sum of two automorphisms in $\bar{S}_2$, say $\bar{h}''_2 = \bar{t}_2 + \bar{t}'_2$.

Set $\bar{\gamma}_1 = \bar{h}'_1 \times \bar{h}'_2$, $\bar{\gamma}_2 = \bar{h}'_1 \times \bar{t}_2$ and $\bar{\gamma}_3 = (-\bar{h}'_1) \times \bar{t}'_2$. Consider the homomorphism $\bar{\gamma} = \bar{\gamma}_1 + \bar{\gamma}_2 + \bar{\gamma}_3$. Then $\bar{\gamma}$ is the sum of three automorphisms $\bar{\gamma}_1, \bar{\gamma}_2$ and $\bar{\gamma}_3$ in $\bar{S}$. Note that for any $x_1 \times x_2 \in \bar{e}\bar{S} = \bar{e}_1\bar{S} \times \bar{e}_2\bar{S}$, we have that $(\bar{h}'_1 \times \bar{h}'_2 + \bar{h}'_1 \times \bar{t}_2 + (-\bar{h}'_1) \times \bar{t}'_2)(x_1 \times x_2) = \bar{h}'_1(x_1) \times \bar{h}_2(x_2) = \bar{h}_1(x_1) \times \bar{h}_2(x_2) = \bar{h}(x_1 \times x_2)$ since $\bar{h}'_1|_{\bar{e}_1\bar{S}} = \bar{h}_1|_{\bar{e}_1\bar{S}}$. This means that $\bar{\gamma}$ is the sum of three automorphisms in $\bar{S}$ and $\bar{\gamma}|_{\bar{e}\bar{S}} = \bar{h}|_{\bar{e}\bar{S}}$. Let us lift the three automorphisms $\bar{\gamma}_i \in \bar{S}$ to automorphisms $\gamma_i \in S$. As M is automorphism-invariant, $\gamma_i(M) \subseteq M$ for $i = 1, 2, 3$. So if we call $\gamma = \gamma_1 + \gamma_2 + \gamma_3$, we get that $\gamma(M) \subseteq M$. Moreover, as $\bar{\gamma}|_{\bar{e}\bar{S}} = \bar{h}|_{\bar{e}\bar{S}}$, there exists a $j \in J(S)$ such that $\gamma|_{eS} = h|_{eS} + j|_{eS}$. Thus, $h|_{eS} = (\gamma - j)|_{eS}$. As $j \in J(S)$, $1 - j$ is an automorphism, and consequently, M is invariant under $1 - j$ and hence under j. We have already seen that M is invariant under γ. Thus, it follows that M is invariant under $\gamma - j$. Call $\varphi = (\gamma - j)|_M$. Then φ is an endomorphism of S such that $\varphi|_{E(N)} = h|_{E(N)}$ and thus $\varphi|_N = f$. This shows that φ extends f, and hence, M is pseudo-injective.

The converse follows straight from Theorem 4.1. □

Theorem 4.4 *Every automorphism-invariant module satisfies the property $C2$.*

Proof Let M be an automorphism-invariant module. Let A be a direct summand of M, and assume $A \cong B$. Since A, being a direct summand of M, is automorphism-invariant, we conclude that B is also automorphism-invariant. Let f be an isomorphism $B \longrightarrow A$; then f is a monomorphism $B \longrightarrow M$. Let $f^{-1} : f(B) \longrightarrow B$ be the inverse of f. As B is automorphism-invariant, f^{-1} extends to $h : A \longrightarrow B$. Set $\varphi = h \circ f$. Then φ is an identity map on B. This shows that f splits; that is, B is a direct summand of M. Hence, M satisfies the property $C2$. □

Lemma 4.5 *If M is an automorphism-invariant module and A and B are closed submodules of M with $A \cap B = 0$, then A and B are relatively injective. Furthermore, for any monomorphism $h : A \to M$ with $A \cap h(A) = 0$, $h(A)$ is closed in M.*

Proof First, let K and T be complements of each other in M. Then, $E(M) = E_1 \oplus E_2$, where $E_1 = E(K)$ and $E_2 = E(T)$. Now let $f : E_1 \to E_2$ be any homomorphism. Then the map $g : E(M) \to E(M)$ defined by $g(x_1 + x_2) = x_1 + x_2 + f(x_1)$ $(x_i \in E_i)$ is an automorphism, so that $f(K) = (g - 1_{E(M)})(K) \subseteq M$. Hence, $f(K) \subseteq E_2 \cap M = T$. Therefore, T is K-injective.

Now if A and B are closed submodules with zero intersection, then, by the preceding argument, A is injective relative to any complement C of A containing B. Therefore, A is B-injective.

Finally, let $h : A \to M$ be a monomorphism with $h(A) \cap A = 0$, and pick any essential closure K of $h(A)$. Since A is K-injective by the preceding arguments, $h^{-1} : h(A) \to A$ extends to a monomorphism $t : K \to A$. Therefore, we must have $h(A) = K$. □

As a consequence, we have the following corollary.

Corollary 4.6 *Let $M = A \oplus B$ be an automorphism-invariant module. Then A is B-injective, and B is A-injective.*

4.2 Nonsingular Automorphism-Invariant Rings

In this section, we will prove a theorem describing right nonsingular automorphism-invariant rings. We have already seen in Therem 3.6 that if M is an automorphism-invariant module, then $M = X \oplus Y$, where X is quasi-injective and Y is a semi-Boolean module. Furthermore, in this case, X and Y are relatively injective modules. In addition, if M is nonsingular, then this result may be strengthened, as we will see in this section.

Definition 4.7 Two modules M and N are said to be orthogonal if no submodule of M is isomorphic to a submodule of N.

Definition 4.8 A module M is said to be a *square module* if there exists a right module N such that $M \cong N^2$ and a submodule N of a module M is called *square-root* in M if N^2 can be embedded in M.

Definition 4.9 A module M is called *square-free* if M contains no nonzero square roots, and M is called *square-full* if every submodule of M contains a nonzero square root in M.

Now we are ready to prove the decomposition theorem for nonsingular automorphism-invariant modules.

Theorem 4.10 *Let M be a nonsingular automorphism-invariant module; then $M = X \oplus Y$, where X is quasi-injective and Y is a square-free module that is orthogonal to X. In this case, for any two submodules D_1 and D_2 of Y with $D_1 \cap D_2 = 0$, $\mathrm{Hom}(D_1, D_2) = 0$ and, consequently, $\mathrm{Hom}(X, Y) = 0 = \mathrm{Hom}(Y, X)$.*

Proof We have already proved the first part of the assertion. Let $f : D_1 \to D_2$ be a nonzero homomorphism. By the nonsingularity, $\mathrm{Ker}(f)$ is closed in D_1, and there is some submodule $L \neq 0$ of D_1 with $\mathrm{Ker}(f) \cap L = 0$. But then, $L \cong f(L) \subseteq D_2$, contradicting the square-freeness of Y. Now the conclusion follows. $\qquad\square$

Theorem 4.11 *The following hold for a nonsingular square-free module M:*

1. *Every closed submodule of M is a fully invariant submodule of M.*
2. *If M is automorphism-invariant, then for any family $\{K_i : i \in I\}$ of closed submodules of M (not necessarily independent), the submodule $\Sigma_{i \in I} K_i$ is automorphism-invariant.*

Proof First, assume that M is square-free and nonsingular. Let K be a closed submodule of M, and let T be a complement in M of K. Suppose that $f \in \mathrm{End}(M)$ with $f(K) \nsubseteq K$. Let $\pi : E(M) \to E(T)$ be the obvious projection with $\mathrm{Ker}(\pi) = E(K)$. Since K is not essential in $f(K) + K$, we have $\pi(K + f(K)) \neq 0$, implying that $\pi(f(K)) \neq 0$, whence $N = T \cap \pi(f(K)) \neq 0$. Then, for $N' = \{x \in K : \pi f(x) \in T\}$, we have $\text{Hom}(N', N) \neq 0$, contradicting the assertion preceding this theorem. This proves (1).

Now assume, furthermore, that M is automorphism-invariant; let $\{K_i : i \in I\}$ be any family of closed submodules of M, and let g be an automorphism of $E(\Sigma_{i \in I} K_i)$. Clearly, g can be extended to an automorphism g' of $E(M)$. Since M is automorphism-invariant, we have $g'(M) \subseteq M$. Then, by (1), $g(K_i) = g'(K_i) \subseteq K_i$ for all $i \in I$. This proves (2). $\qquad\square$

Theorem 4.12 *If R is a right nonsingular right automorphism-invariant ring, then $R \cong S \times T$, where S and T are rings with the following properties:*

1. *S is a right self-injective ring.*
2. *T_T is square-free.*
3. *Any sum of closed right ideals of T is a two-sided ideal that is automorphism-invariant as a right T-module.*
4. *For any prime ideal P of T that is not essential in T_T, T/P is a division ring.*

Proof (1) and (2) follow straight from Theorem 3.6. Also, (3) follows from Theorem 4.11.

We now prove (4): Let P be a prime ideal of T that is not essential as a right ideal. Take a complement N of P in T_T. If N were not uniform, there would be two nonzero closed right ideals in N, say X and Y with $X \cap Y = 0$. They would then be ideals by the preceding argument. But this would contradict the primeness of P. So N is a uniform right ideal of T. Also note that P is a closed submodule of T_T, because if P' is any essential extension of P, we have $P'N = 0$, implying that $P' = P$. So P is closed in T_T, and hence, it is a complement in T_T of N. Since $(N \oplus P)/P$ is essential in T/P, this implies that the ring T/P is right uniform. Furthermore, N is a nonsingular uniform automorphism-invariant T-module so that every nonzero homomorphism between any two submodules is an isomorphism between essential submodules, and thus, it extends to an automorphism of N. Therefore, N is a quasi-injective uniform nonsingular T/P-module, and thus, its endomorphism ring is a division ring. Since T/P essentially contains the nonsingular right ideal $(N \oplus P)/P$, it is now a prime right uniform and right nonsingular ring (hence a prime right Goldie ring) with the quasi-injective essential right ideal $(N \oplus P)/P$. But then $(N \oplus P)/P$ is injective, implying that $P \oplus N = T$. In fact, since $\text{End}_{T/P}(N)$ is a division ring, T/P is a division ring. In particular, N is a simple right ideal, and P is a maximal right ideal of T. $\qquad\square$

Theorem 4.13 *If R is a prime right nonsingular, right automorphism-invariant ring, then R is right self-injective.*

Proof By Theorem 4.12 and primeness, it suffices to look at the case when R_R is square-free: If R_R were not uniform, there would be two closed nonzero right ideals A and B with $A \cap B = 0$. But then A and B would be ideals, whence $AB = 0$, contradicting primeness. So R_R is uniform, nonsingular and automorphism-invariant. Now it follows in the same way as in the proof of Theorem 4.12 that R is right self-injective. □

In Example 2.12, we have already seen that the conclusion of Theorem 4.13 fails if we take a semiprime ring instead of a prime one.

Corollary 4.14 *A simple right automorphism-invariant ring is right self-injective.*

4.3 When Is an Automorphism-Invariant Module a Quasi-Injective Module

In this section, we will apply our previously obtained results to characterize when automorphism-invariant modules are quasi-injective modules. We have already seen that if M is right R-module such that $\mathrm{End}(M)$ has no factor isomorphic to $\mathbb{F}_2$, then M is quasi-injective if and only M is automorphism-invariant. Now we will see some more results in the same direction.

Lemma 4.15 *Let R be any ring, and let S be a subring of its center Z(R). If $\mathbb{F}_2$ does not admit a structure of right S-module, then for any right R-module M, the endomorphism ring $\mathrm{End}(M)$ has no factor isomorphic to $\mathbb{F}_2$.*

Proof Assume to the contrary that there exists a ring homomorphism $\psi \colon \mathrm{End}_R(M) \to \mathbb{F}_2$. Now, define a map $\varphi \colon S \to \mathrm{End}_R(M)$ by the rule $\varphi(r) = \varphi_r$, for each $r \in S$, where $\varphi_r \colon M \to M$ is given as $\varphi_r(m) = mr$. Clearly, φ is a ring homomorphism since $S \subseteq Z(R)$, and so the composition $\varphi \circ f$ gives a nonzero ring homomorphism from S to $\mathbb{F}_2$, yielding a contradiction to the assumption that $\mathbb{F}_2$ does not admit a structure of right S-module. □

This leads us to the following.

Theorem 4.16 *Let A be an algebra over a field $\mathbb{F}$ with more than two elements. Then any right A-module M is automorphism-invariant if and only if M is quasi-injective.*

Proof Let M be an automorphism-invariant right A-module. Since A is an algebra over a field $\mathbb{F}$ with more than two elements, by Lemma 4.15, it follows that $\mathbb{F}_2$ does not admit a structure of right $Z(A)$-module, and therefore, $\mathrm{End}(M)$ has no factor isomorphic to $\mathbb{F}_2$. Now, by Theorem 5.39, M must be quasi-injective. The converse is obvious.   □

As a consequence, we have the following:

Corollary 4.17 *Let R be any algebra over a field $\mathbb{F}$ with more than two elements. Then R is of right invariant module type if and only if every indecomposable right R-module is automorphism-invariant.*

Corollary 4.18 *If A is an algebra over a field $\mathbb{F}$ with more than two elements such that A is automorphism-invariant as a right A-module, then A is right self-injective.*

It is well known that a group ring $R[G]$ is right self-injective if and only if R is right self-injective and G is finite. Thus, in particular, we have the following:

Corollary 4.19 *Let $K[G]$ be automorphism-invariant, where K is a field with more than two elements. Then G must be finite.*

Lemma 4.20 *If a uniform module is automorphism-invariant, then it must be quasi-injective.*

Proof Suppose M is uniform and automorphism-invariant. Since $E(M)$ is indecomposable, $\mathrm{End}(E(M))$ is a local ring. Now if we take any endomorphism $f \in \mathrm{End}(E(M))$, then as $\mathrm{End}(E(M))$ is a local ring, either f is an automorphism or $1 - f$ is an automorphism. Clearly, in any case, then M will be invariant under f. This shows M is quasi-injective.   □

Note that if M is an automorphism-invariant module of finite Goldie dimension, then $M = \oplus_{i=1}^{n} M_i$ is a direct sum of indecomposable modules also having finite Goldie dimension. Moreover, a finite direct sum of automorphism-invariant modules is again automorphism-invariant if and only if the direct summands are relatively injective. Thus, the module M is quasi-injective if and only if so is each M_i. This means that the question of whether an automorphism-invariant module with finite Goldie dimension is quasi-injective reduces to study when an indecomposable automorphism-invariant module having finite Goldie dimension is quasi-injective. In the following example, we see that an indecomposable automorphism invariant module having finite Goldie dimension need not be quasi-injective.

Example 4.21 Let $R = \begin{bmatrix} \mathbb{F} & \mathbb{F} & \mathbb{F} \\ 0 & \mathbb{F} & 0 \\ 0 & 0 & \mathbb{F} \end{bmatrix}$ where $\mathbb{F}$ is a field of order 2.

We know that R is a left serial ring. Note that $e_{11}R$ is a local module, $e_{12}\mathbb{F} \cong e_{22}R$, $e_{13}\mathbb{F} \cong e_{33}R$ and $e_{11}J(R) = e_{12}\mathbb{F} \oplus e_{13}\mathbb{F}$, a direct sum of two minimal right ideals. So the injective envelope of $e_{11}R$ is $E(e_{11}R) = E_1 \oplus E_2$, where $E_1 = E(e_{12}\mathbb{F})$ and $E_2 = E(e_{13}\mathbb{F})$. Since R is a finite-dimensional $\mathbb{F}_2$-algebra, $e_{11}R$ is an Artinian right R-module, and hence, it has finite Goldie dimension.

Now set $A = r(e_{12}\mathbb{F})$. Then $A = e_{13}\mathbb{F} + e_{33}\mathbb{F}$. Thus, $\overline{R} = R/A \cong \begin{bmatrix} \mathbb{F} & \mathbb{F} \\ 0 & \mathbb{F} \end{bmatrix} = S$. Denote the first row of S by S_1. It may be checked that S_1 is injective. As $\mathbb{F}$ has only two elements, S_1 has only two endomorphisms, zero and the identity. Take the pre-image L_1 of S_1 in $\overline{R}$. It is uniserial with composition length 2, and $e_{12}\mathbb{F}$ naturally embeds in L_1. There is no mapping of $e_{13}\mathbb{F}$ into L_1. It follows that L_1 is $e_{11}R$-injective and $e_{12}\mathbb{F}$-injective. As $e_{22}R \cong e_{12}\mathbb{F}$, L_1 is $e_{22}R$-injective. There is no map from $e_{33}R$ into L_1, so it is also $e_{33}R$-injective. Hence, L_1 is injective. Thus, $E_1 = L_1$, and its ring of endomorphisms has only two elements.

If $B = r(e_{13}\mathbb{F})$, then $B = e_{12}\mathbb{F} + e_{22}\mathbb{F}$. Thus, $R/B \cong \begin{bmatrix} \mathbb{F} & \mathbb{F} \\ 0 & \mathbb{F} \end{bmatrix}$. The pre-image of S_1 in R/B is L_2, which is uniserial and injective. We have $E_2 \cong L_2$, and its ring of endomorphism has only two elements.

Note that $e_{11}R$ has all its composition factors non-isomorphic, both L_1 and L_2 have composition length 2 with $L_1/(L_1J(R)) \cong (e_{11}R)/(e_{11}J(R))$, $L_1J(R) \cong e_{22}R$, $L_2/(L_2J(R)) \cong (e_{11}R)/(e_{11}J(R))$ and $L_2J(R) \cong e_{33}R$. Thus, L_1, L_2 have isomorphic tops but non-isomorphic socles.

Suppose there exists a nonzero mapping $\sigma : L_1 \to L_2$. Then $\sigma(L_1) = L_2J(R)$. Thus, $(e_{11}R)/(e_{11}J(R)) \cong e_{33}R$, which is a contradiction. Therefore, there is no nonzero map between L_1 and L_2.

Hence, the only automorphism of $L_1 \oplus L_2$ is the identity. So $e_{11}R$ is trivially automorphism-invariant, but it is not uniform. Then, clearly, $e_{11}R$ is not quasi-injective, as an indecomposable quasi-injective module must be uniform. Thus, we have an example of an indecomposable module with finite Goldie dimension that is automorphism-invariant but not quasi-injective.

Example 4.22 Let $A = \mathbb{F}_2[x]$ and

$$R = \begin{bmatrix} A/(x) & 0 \\ A/(x) & A/(x^2) \end{bmatrix}.$$

Let $M = \begin{bmatrix} 0 & 0 \\ A/(x) & A/(x^2) \end{bmatrix}$. As $M = e_{22}R$, where e_{22} is a primitive idempotent, M is an indecomposable right R-module. Note that M has two simple submodules $S_1 = \begin{bmatrix} 0 & 0 \\ A/(x) & 0 \end{bmatrix}$ and $S_2 = \begin{bmatrix} 0 & 0 \\ 0 & (x)/(x^2) \end{bmatrix}$ such that $S_1 \oplus S_2$ is essential in M.

Clearly, R is a finite-dimensional $\mathbb{F}_2$-algebra. Let $\mathcal{X}$ be the class of injective right R-modules. It is not difficult to check that M is automorphism-invariant. But M is not quasi-injective, as M is not uniform. This gives another example of an indecomposable module with finite Goldie dimension that is automorphism-invariant but not quasi-injective.

We will begin our characterization of indecomposable automorphism-invariant modules by proving the following theorem.

Theorem 4.23 *Let M be an indecomposable automorphism-invariant module with finite Goldie dimension such that M is not quasi-injective. Then,*

1. *$\text{End}(M)/J(\text{End}(M)) \cong \mathbb{F}_2$.*
2. *There exists a finite set of non-isomorphic indecomposable injective modules $\{E_i\}_{i=1}^{n}$, with $n \geq 2$, such that $E(M) = \oplus_{i=1}^{n} E_i$ and $\text{End}(E_i)/J(\text{End}(E_i)) \cong \mathbb{F}_2$ for every $i = 1, \ldots, n$.*

Proof By Theorem 2.32, $\text{End}(M)/J(\text{End}(M)) \cong \mathbb{F}_2$. On the other hand, we know from Theorem 3.6 that $M = N \oplus L$ is the direct sum of a quasi-injective module L and a semi-Boolean module N. As M is indecomposable and non quasi-injective, this means that $M = N$. Call $E = E(M)$ the injective envelope of M. Since M has finite Goldie dimension, E must be a finite direct sum of indecomposable injective modules. But again, the proof of Theorem 3.6 shows that $\text{End}(E)/J(\text{End}(E))$ is a Boolean ring, and therefore, E is square free. So $E = E_1 \oplus \cdots \oplus E_n$ for some indecomposable injective modules E_i satisfying that E_i is non-isomorphic to E_j for any $j \neq i$. Moreover, Theorem 3.7 shows that $n \geq 2$. Finally, as each E_i is an indecomposable injective module, $\text{End}(E_i)/J(\text{End}(E_i))$ is a division ring for every i. And, as $\text{End}(E)/J(\text{End}(E))$ is a Boolean ring, this means that $\text{End}(E_i)/J(\text{End}(E_i)) \cong \mathbb{F}_2$ for each $i = 1, \ldots, n$. $\qquad \square$

In particular, if M is a finitely cogenerated module, its socle $\text{Soc}(M)$ is finitely generated and essential in M. So we get the following corollary.

Corollary 4.24 *Let M be an indecomposable finitely cogenerated automorphism-invariant module that is not quasi-injective. If we write $\text{Soc}(M) = \oplus_{i=1}^{n} C_i$ as a direct sum of indecomposable modules, then*

1. $n \geq 2$.
2. $\mathrm{End}(M)/J(\mathrm{End}(M)) \cong \mathbb{F}_2$.
3. $\mathrm{End}(C_i) \cong \mathbb{F}_2$ *for every* $i \in I$.
4. $C_i \not\cong C_j$ *if* $i \neq j$.

Proof Let us note that $E(M) = \oplus_{i=1}^{n} E(C_i)$. Therefore, the result is an immediate consequence of the preceding theorem. □

Our next step will be to extend the preceding corollary to automorphism-invariant modules over right bounded right Noetherian rings.

Definition 4.25 A ring R is called right bounded if each essential right ideal of R contains a two-sided ideal that is essential as a right ideal.

A right Noetherian ring R is right bounded if and only if each essential right ideal of R contains a nonzero two-sided ideal [69].

Definition 4.26 A ring R is called a right FBN ring if R is a right Noetherian ring such that R/P is right bounded for each prime ideal P of R.

It is well known that over a right bounded right Noetherian ring R, any indecomposable injective right module is of the form $E(U_R)$, where U_R is a uniform right submodule of R/P for some prime ideal P. Furthermore, if R is right FBN, and U_R, U'_R are two nonzero uniform right submodules of R/P, then $E(U) \cong E(U')$ [41, p. 163].

We now proceed to give a sufficient condition for an automorphism-invariant module over a right bounded right Noetherian ring to be quasi-injective. First, we have the following technical proposition.

Proposition 4.27 *Let R be a right bounded right Noetherian ring and M, a non-quasi-injective, semi-Boolean, automorphism-invariant module. Then there exists a set $\{C_i\}_{i \in I}$ of non-isomorphic simple right R-modules with $|I| \geq 2$ such that $\mathrm{Soc}(M) = \oplus_{i \in I} C_i$ is essential in M and $\mathrm{End}(C_i) \cong \mathbb{F}_2$ for every $i \in I$. Moreover, if R is a commutative ring, then $|C_i| = 2$ for every $i \in I$.*

Proof Let us first note that as R is a right bounded ring, there exist prime ideals $\{P_i\}_{i \in I}$ and nonzero uniform right ideals $\{U_i\}_{i \in I}$ such that $E(M) = \oplus_{i \in I} E(U_i)$ and each U_i is a submodule of R/P_i. Moreover, $E(M)$ is also a semi-Boolean module, and thus, $\mathrm{End}(E(M))/J(\mathrm{End}(E(M)))$ is square-free. It follows then that $E(U_i) \not\cong E(U_j)$ if $i \neq j$. Furthermore, each $\mathrm{End}(E(U_i))/J(\mathrm{End}(E(U_i)))$ is a division ring. Thus, $\mathrm{End}(E(U_i))/J(\mathrm{End}(E(U_i))) \cong \mathbb{F}_2$ for every $i \in I$, again because $\mathrm{End}(E(M))/J(\mathrm{End}(E(M)))$

is a Boolean ring. Finally, as $\text{End}(E(M))$ must have a homomorphic image isomorphic to $\mathbb{F}_2 \times \mathbb{F}_2$ by Theorem 3.7, we deduce that $|I| \geq 2$.

Fix now an index $i \in I$, and call $P = P_i, U = U_i, Q = E(U)$ and $T = R/P$. Let us choose a nonzero right ideal $V \subseteq U$. Call $L = \{v \in V \mid r_T(v) \cap V \neq 0\}$, where $r_T(v)$ denotes the right annihilator of v in T.

We first show that any element in L is nilpotent. Let us choose a nonzero element $v \in V$.

We have an ascending chain of right ideals

$$r_T(v) \subseteq r_T(v^2) \subseteq \cdots \subseteq r_T(v^m) \subseteq \cdots$$

So there exists an m_0 such that $r_T(v^{m_0}) = r_T(v^{m_0+m})$ for each $m \in \mathbb{N}$, since R is right Noetherian. Assume that $v^{m_0} \neq 0$. As $v^{m_0} T \subseteq V$ and $0 \neq r_T(v) \cap V$ is an essential submodule of V, since V is uniform, there exists a $t \in T$ such that $0 \neq v^{m_0} t \in r_T(v)$. And thus, $v^{m_0+1} t = 0$. But, as $r_T(v^{m_0}) = r_T(v^{m_0+1})$, this means that $v^{m_0} t = 0$, a contradiction that shows that any element in L is nilpotent.

Therefore, L is a nil subset of $T = R/P$. Now, it is easy to check that L is multiplicatively closed. So L is nilpotent by [41, theorem 6.21]. We claim that $L^2 = 0$. Let m_0 the biggest integer such that there exist elements $v_1, \ldots, v_{m_0} \in L$ such that $v_{m_0} \cdots \cdot v_1 \neq 0$, and assume on the contrary that $m_0 \geq 2$. Then, for any $t \in T$, we have that $v_1 t v_2 \in L$, and thus, $v_{m_0} \cdots \cdot (v_1 t v_2) v_1$ is the product of $m_0 + 1$ elements in L, and so it is 0. It follows that $v_{m_0} \cdots \cdot v_1 T v_2 \cdot v_1 = 0$, and thus, $v_{m_0} \cdots \cdot v_1 = 0$ or $v_2 v_1 = 0$ since T is a prime ring. In either case, we deduce that $v_{m_0} \cdots \cdot v_1 = 0$ since $m_0 \geq 2$. A contradiction that proves our claim.

So we have that $L^2 = 0$. Let us distinguish two possibilities.

Case 1. If L is a right ideal of T, then $vT \subseteq L$ for any $v \in L$, and thus $vTv = 0$ as $L^2 = 0$. Again, as T is prime, we deduce that $v = 0$, and thus, $L = 0$. But then, the left multiplication $f_v : V \to V$ by v is a monomorphism for every nonzero $v \in V$ that extends to an isomorphism $g_v : E(V) \to E(V)$. We claim that V has only two elements. Let us choose two nonzero elements $v, v' \in V$ and assume that they are not equal. Then $g_v, g_{v'}$ are two automorphisms of $E(V)$, and thus, $g_v - g_{v'} \in J(\text{End}(E(V)))$ since $\text{End}(E(V))/J(\text{End}(E(V)) \cong \mathbb{F}_2$. Therefore, $g_v - g_{v'}$ has essential kernel, and so $v - v' \in L = 0$, and we get that $v = v'$. This means that V has only two elements, and thus, it is a simple submodule of T satisfying that $Q = E(V)$.

Case 2. Assume that L is not a right ideal of T. This means that there exists a $v_0 \in L$ and a $t_0 \in T$ such that $v_0 t_0 \notin L$. Therefore, $r_T(v_0 t_0) \cap V = 0$, and this means that $r_T((v_0 t_0)^m) \cap V = 0$ for each $m \geq 1$, as V is uniform.

Call $v_1 = v_0 t_0$, and let $f_{v_1^m} : V \to V$ be the left multiplication by v_1^m. Then $f_{v_1^m}$ is a monomorphism for each $m \geq 1$, and reasoning as in Case 1, we deduce that $v_1^m - v_1^{m'} \in L$ for each $m, m' \geq 1$. In particular, $v := v_1 - v_1^3 \in L$. But then $v_1 v_0 - v_1^3 v_0 = v v_0 = 0$ since $v, v_0 \in L$. And multiplying on the right by t_0, we get that $e = v_1^2$ is a nonzero idempotent of V. As U was uniform, we deduce that $U = V$ is a direct summand of T.

Therefore, either we are in Case 1 for some nonzero submodule V of W and we get that Q is the injective envelope of the simple module V having only two elements, or otherwise, we get that any nonzero submodule of U equals U. Therefore, U is a simple right ideal of T.

Finally, note that if R is a commutative ring, then L is a right ideal of T, and thus, we are always in Case 1. This completes the proof. $\square$

Remark 4.28 Assume that the ring R in the preceding proposition is right FBN. Then, for any simple right R-module C, we have that $r_R(C)$ is a maximal two-sided ideal of R and $R/r_R(C)$ is a simple Artinian ring (see [41, theorem 9.10]). Moreover, $R/r_R(C) \cong C_R^n$ for some $n \geq 1$ as a right R-module. Therefore,

$$R/r_R(C) \cong \mathrm{End}_R(R/r_R(C)) \cong \mathrm{End}_R(C^n) \cong M_n(\mathbb{F}_2).$$

This means that, under this identification, C becomes a simple right ideal of the full matrix ring $M_n(\mathbb{F}_2)$, and thus, $|C| = 2^n$, where n is the Goldie dimension of $R/r_R(C)$.

In particular, when M is an indecomposable module, we get the following:

Corollary 4.29 *Let R be a right bounded right Noetherian ring, and let M be an indecomposable automorphism-invariant right R-module that is not quasi-injective. Then there exists a set $\{C_i\}_{i \in I}$ of non-isomorphic simple right R-modules with $|I| \geq 2$ such that $\mathrm{Soc}(M) = \oplus_{i \in I} C_i$ is essential in M and $\mathrm{End}(C_i) \cong \mathbb{F}_2$ for each $i \in I$.*

Proof We know from Theorem 3.6 that $M = N \oplus L$, where N is a semi-Boolean module, any element in $\mathrm{End}(E(L))/J(\mathrm{End}(E(L)))$ is the sum of two units and moreover, $\mathrm{End}(E(M)/J(\mathrm{End}(E(M)) = \mathrm{End}(E(N))/J(\mathrm{End}(E(N)) \times \mathrm{End}(E(L))/J(\mathrm{End}(E(L)))$. As we are assuming that M is indecomposable and not quasi-injective, we must have $N = M$. So the result follows from the preceding proposition. $\square$

Another consequence of Proposition 4.27 is the following corollary, which characterizes when every automorphism-invariant module over a commutative Noetherian ring is quasi-injective.

Corollary 4.30 *Let R be a commutative Noetherian ring with no homomorphic images isomorphic to $\mathbb{F}_2 \times \mathbb{F}_2$. Then any automorphism-invariant R-module is quasi-injective.*

Proof Assume on the contrary that M is an automorphism-invariant R-module which is not quasi-injective. Again, we know from Theorem 3.6 that $M = N \oplus L$, where N is a semi-Boolean module, any element in $\text{End}(E(L))/J(\text{End}(E(L)))$ is the sum of two units and $\text{End}(E(M))/J(\text{End}(E(M)) = \text{End}(E(N))/J(\text{End}(E(N)) \times \text{End}(E(L))/J(\text{End}(E(L))$. So N must be also an automorphism-invariant module that is not quasi-injective. Applying now Proposition 4.27, we deduce that there exists a set $\{C_i\}_{i \in I}$ of non-isomorphic simple right R-modules with $|I| \geq 2$ such that $\text{Soc}(N) = \oplus_{i \in I} C_i$ is essential in N and C_i has only two elements for every $i \in I$.

Write $C_i = \{0, c_i\}$ for every $i \in I$. Let $p_i \colon R \to C_i$ be the homomorphism defined by $p_i(1) = c_i$ and call $N_i = \text{Ker}(p_i)$. Then N_i is a maximal ideal of R.

Finally, as $|I| \geq 2$, there exist least two different indexes $i_1, i_2 \in I$. Define then $p \colon R \to R/N_{i_1} \times R/N_{i_2}$ by $p(1) = p_{i_1}(1) \times p_{i_2}(1)$. Clearly, p is a ring homomorphism, and it is easy to check that $R/N_{i_1} \times R/N_{i_2}$ is isomorphic to the ring $\mathbb{F}_2 \times \mathbb{F}_2$. Let us check that p is surjective. As N_{i_1}, N_{i_2} are different maximal deals, there exist elements $r_{i_1} \in N_{i_1} \backslash N_{i_2}$ and $r_{i_2} \in N_{i_2} \backslash N_{i_1}$. And this means that $p(r_{i_1}) = c_{i_1} \times 0$ and $p(r_{i_2}) = 0 \times c_{i_2}$. Therefore, p is surjective. This yields a contradiction to our assumption that R has no homomorphic images isomorphic to $\mathbb{F}_2 \times \mathbb{F}_2$. Thus, it follows that any automorphism-invariant R-module must be quasi-injective. □

In particular, for a commutative Noetherian local ring, we conclude the following.

Corollary 4.31 *Over a commutative Noetherian local ring, any automorphism-invariant module is quasi-injective.*

Proof If R has a homomorphic image isomorphic to $\mathbb{F}_2 \times \mathbb{F}_2$, then it has at last two distinct maximal ideals, and so it cannot be local. This shows that if R is a commutative Noetherian local ring, then it has no homomorphic images isomorphic to $\mathbb{F}_2 \times \mathbb{F}_2$, and consequently, any automorphism-invariant R-module is quasi-injective. □

Remark 4.32 Let us note that any finite-dimensional algebra over a field K is a left and right FBN ring by [41, proposition 9.1(a)]. Therefore, Examples 4.21

and 4.22 show that a finitely generated automorphism-invariant module over a (two-sided) FBN ring R does not need to be quasi-injective.

Our next step will be to show that any automorphism-invariant module over a commutative Noetherian ring R is quasi-injective. In order to prove it, we will first show that being automorphism-invariant is a local property for modules over commutative Noetherian rings.

Proposition 4.33 *Let R be a commutative Noetherian ring, and let M be an R-module. The following are equivalent.*

1. *M is automorphism-invariant.*
2. *$M_{\mathfrak{p}}$ is an automorphism-invariant $R_{\mathfrak{p}}$-module for every prime ideal $\mathfrak{p}$.*
3. *$M_{\mathfrak{m}}$ is an automorphism-invariant $R_{\mathfrak{m}}$-module for every maximal ideal $\mathfrak{m}$.*

Proof (1) $\Rightarrow$ (2). Let $\mathfrak{p}$ be a prime ideal, and assume that M is an automorphism-invariant module. Call $E = E(M)$. As the torsion submodules in R-Mod for the multiplicative set $S_{\mathfrak{p}} = R \setminus \mathfrak{p}$ are closed under injective envelopes (see e.g. [98, proposition 4.5 (i)]), we get that $E = t(E) \oplus E'$, where $t(E)$ is the torsion submodule of E and $E' = E(M_{\mathfrak{p}})$ is the injective envelope of $M_{\mathfrak{p}}$. Now let $f : E' \to E'$ be an automorphism of E'. Then the diagonal homomorphism $1_{t(M)} \oplus f : E \to E$ is an isomorphism of E, and thus, $(1_{t(M)} \oplus f)(M) \subseteq M$ by hypothesis. And this means that $f(M_{\mathfrak{p}}) = (1_{t(M)} \oplus f)_{\mathfrak{p}}(M_{\mathfrak{p}}) \subseteq M_{\mathfrak{p}}$. So $M_{\mathfrak{p}}$ is automorphism-invariant.

(2) $\Rightarrow$ (3). This is trivial.

(3) $\Rightarrow$ (1). Now fix an automorphism $f : E(M) \to E(M)$, and choose a maximal ideal $\mathfrak{m}$ of R. The localization $f_{\mathfrak{m}}$ of f at $\mathfrak{m}$ is also an isomorphism. And localizing at $\mathfrak{m}$ the short exact sequence

$$ M \xrightarrow{f|_M} M + f(M) \to (M + f(M))/f(M) \to 0, $$

we obtain the short exact sequence

$$ M_{\mathfrak{m}} \xrightarrow{f_{\mathfrak{m}}|_{M_{\mathfrak{m}}}} M_{\mathfrak{m}} + f(M)_{\mathfrak{m}} = M_{\mathfrak{m}} + f_{\mathfrak{m}}(M_{\mathfrak{m}}) $$
$$ \to (M_{\mathfrak{m}} + f_{\mathfrak{m}}(M_{\mathfrak{m}}))/f_{\mathfrak{m}}(M_{\mathfrak{m}}) \to 0. $$

As $M_{\mathfrak{m}}$ is automorphism-invariant, we deduce that $(M_{\mathfrak{m}} + f_{\mathfrak{m}}(M_{\mathfrak{m}}))/f_{\mathfrak{m}}(M_{\mathfrak{m}}) = 0$. Therefore, $(M + f(M)/f(M))_{\mathfrak{m}} = 0$ for every maximal ideal $\mathfrak{m}$, and this means that $(M + f(M))/f(M) = 0$. Thus, M is invariant under f. $\square$

Bearing in mind that a module M is quasi-injective if and only if it is invariant under any endomorphism of its injective envelope, the preceding arguments can be easily adapted to prove:

Proposition 4.34 *Let R be a commutative Noetherian ring, and let M be an R-module. The following are equivalent.*

1. *M is quasi-injective.*
2. *$M_\mathfrak{p}$ is a quasi-injective $R_\mathfrak{p}$-module for every prime ideal $\mathfrak{p}$.*
3. *$M_\mathfrak{m}$ is a quasi-injective $R_\mathfrak{m}$-module for every maximal ideal $\mathfrak{m}$.*

We can now extend Corollaries 4.30 and 4.31 and prove the following.

Theorem 4.35 *Let R be a commutative Noetherian ring. Then any automorphism-invariant R-module is quasi-injective.*

Proof Let M be an automorphism-invariant R-module. Then $M_\mathfrak{p}$ is an automorphism-invariant $R_\mathfrak{p}$-module for every prime ideal $\mathfrak{p}$ by Proposition 4.33. As each $R_\mathfrak{p}$ is local, we deduce from Corollary 4.31 that $M_\mathfrak{p}$ is quasi-injective for every prime ideal $\mathfrak{p}$. And therefore, M is quasi-injective by Proposition 4.34.   □

In [13], it was asked whether for an abelian group the notions of automorphism-invariance and quasi-injectivity coincide. The preceding theorem answers this question by taking ring R to be $\mathbb{Z}$, and we state it formally in what follows.

Corollary 4.36 *For abelian groups, the notions of automorphism-invariance and quasi-injectivity coincide.*

Remarks 4.37 We note here that the assumptions in the preceding theorem are essential in getting the conclusion.

1. In [100], Teply constructed an example of a non-finitely generated automorphism-invariant module over the commutative ring $\mathbb{Z}[x_1, x_2, \ldots, x_n, \ldots]$ that is not quasi-injective. Therefore, we cannot expect to extend our previous theorem to modules over non-Noetherian commutative rings.
2. On the other hand, the ring R of all eventually constant sequences of elements of the field $\mathbb{F}_2$ is a commutative Boolean ring that is automorphism-invariant as a module over itself, but it is not self-injective. This shows that we cannot drop the Noetherian condition from the hypotheses of the preceding theorem, even if we assume that the ring is Boolean and the module is finitely generated and nonsingular.

It is straightforward to check that being automorphism-invariant is a Morita-invariant property. Therefore, the statement in Theorem 4.35 is also valid if we replace "commutative Noetherian ring R" with "a full matrix ring $\mathbb{M}_n(R)$ over a commutative Noetherian ring R."

Next, we proceed to show that under certain conditions, a decomposition of injective envelope $E(M)$ of an automorphism-invariant module M induces a natural decomposition of M. We will denote the identity automorphism on any module M by I_M.

Lemma 4.38 *Let M be an automorphism-invariant right module over any ring R. If $E(M) = E_1 \oplus E_2$ and $\pi_1 \colon E(M) \to E_1$ is an associated projection, then $M_1 = \pi_1(M)$ is also automorphism-invariant.*

Proof Let $E(M) = E_1 \oplus E_2$ and $M_1 = \pi_1(M)$, where $\pi_1 \colon E(M) \to E_1$ is a projection with E_2 as its kernel. Let σ_1 be an automorphism of E_1 and $x_1 \in M_1$. For some $x \in M$, and $x_2 \in E_2$, we have $x = x_1 + x_2$. Now $\sigma = \sigma_1 \oplus I_{E_2}$ is an automorphism of E. Thus, $\sigma(x) = \sigma_1(x_1) + x_2 \in M$, which gives $\sigma_1(x_1) \in M_1$. Hence, M_1 is automorphism-invariant. $\qquad\square$

Lemma 4.39 *Let M be an automorphism-invariant right module over any ring R. Let $E(M) = E_1 \oplus E_2$ such that there exists an automorphism σ_1 of E_1 such that $I_{E_1} - \sigma_1$ is also an automorphism of E_1. Then*

$$M = (M \cap E_1) \oplus (M \cap E_2).$$

Proof Set $E = E(M)$. Set $I_E = I_{E_1} \oplus I_{E_2}$, and $\sigma = \sigma_1 \oplus I_{E_2}$. Clearly, both I_E and σ are automorphisms of E. Since M is assumed to be an automorphism-invariant module, M is invariant under automorphisms I_E and σ. Consequently, M is invariant under $I_E - \sigma$, too. Note that $I_E - \sigma = (I_{E_1} - \sigma_1) \oplus 0$. Thus, $(I_E - \sigma)(M) = (I_{E_1} - \sigma_1)(M) \subseteq M$. Let $\pi_1 \colon E \to E_1$ and $\pi_2 \colon E \to E_2$ be the canonical projections. Set $M_1 = \pi_1(M)$ and $M_2 = \pi_2(M)$. Now $M \cap E_1 \subseteq M_1$ and $M \cap E_2 \subseteq M_2$.

Let $0 \neq u_1 \in E_1$. For some $r \in R$, $0 \neq u_1 r \in M$ and, thus, $u_1 r \in M_1$. Therefore, $M_1 \subseteq_e E_1$. By Lemma 4.38, M_1 is automorphism-invariant. Thus, $M_1 = (I_{E_1} - \sigma_1)^{-1}(M_1)$. Let $x_1 \in M_1$. Then we have, for some $x \in M$, $x = x_1 + x_2$, $x_2 \in E_2$. Now, as $I_{E_1} - \sigma_1$ is an automorphism on E_1, there exists an element $y_1 \in E_1$ such that $(I_{E_1} - \sigma_1)(y_1) = x_1$, which gives $y_1 \in (I_{E_1} - \sigma_1)^{-1}(M_1) = M_1$. This yields an element $y \in M$ such that $y = y_1 + y_2$ for some $y_2 \in E_2$. We get $(I_E - \sigma)(y) = (I_{E_1} - \sigma_1)(y_1) = x_1$. Thus, $x_1 \in (I_E - \sigma)(M)$. As $(I_E - \sigma)(M) \subseteq M$, we get $x_1 \in M$. Hence, $M_1 \subseteq M$.

Now, let $u_2 \in M_2$ be an arbitrary element. For some $u_1 \in M_1$, we have $u = u_1 + u_2 \in M$. But we have shown in the previous paragraph that $M_1 \subseteq M$, so $u_1 \in M$. Therefore, $u_2 = u - u_1 \in M$. Hence, $M_2 \subseteq M$. This gives $M_1 \oplus M_2 \subseteq M$ and, hence, $M = M_1 \oplus M_2$. Thus, $M = (M \cap E_1) \oplus (M \cap E_2)$. $\qquad\square$

Theorem 4.40 *Let M be a right module over any ring R such that every summand E_1 of $E(M)$ admits an automorphism σ_1 such that $I_{E_1} - \sigma_1$ is also an automorphism of E_1; then M is automorphism-invariant if and only if M is quasi-injective.*

Proof Let M be automorphism-invariant. Set $E = E(M)$. Suppose every summand E_1 of E admits an automorphism σ_1 such that $I_{E_1} - \sigma_1$ is also an automorphism of E_1.

Let $\sigma \in \mathrm{End}(E)$ be an arbitrary element. Since $\mathrm{End}(E)$ is a clean ring, $\sigma = \alpha + \beta$, where α is an idempotent and β is an automorphism.

Let $E_1 = \alpha E$, and $E_2 = (1 - \alpha)E$. Then $E = E_1 \oplus E_2$. By Lemma 4.39, we have $M = M_1 \oplus M_2$ where $M_1 = M \cap E_1$, $M_2 = M \cap E_2$.

Then, clearly, $\alpha(M) \subseteq M$. Since M is automorphism-invariant, $\beta(M) \subseteq M$. Thus, $\sigma(M) \subseteq M$. Hence, M is quasi-injective.

The converse is obvious. $\qquad\square$

Lemma 4.41 *Let M be an automorphism-invariant right module over any ring R. If $E(M) = E_1 \oplus E_2 \oplus E_3$, where $E_1 \cong E_2$, then*

$$M = (M \cap E_1) \oplus (M \cap E_2) \oplus (M \cap E_3).$$

Proof Set $E(M) = E$. Let $E = E_1 \oplus E_2 \oplus E_3$. Let $\sigma : E_1 \to E_2$ be an isomorphism and let $\pi_1 : E \to E_1$, $\pi_2 : E \to E_2$ and $\pi_3 : E \to E_3$ be the canonical projections. Then $M \cap E_1 \subseteq \pi_1(M)$, $M \cap E_2 \subseteq \pi_2(M)$ and $M \cap E_3 \subseteq \pi_3(M)$.

Let $\eta = \sigma^{-1}$. Consider the map $\lambda_1 : E \to E$ given by $\lambda_1(x_1, x_2, x_3) = (x_1, \sigma(x_1) + x_2, x_3)$. Clearly, λ_1 is an automorphism of E. Since M is automorphism-invariant, M is invariant under λ_1 and I_E. Consequently, M is invariant under $\lambda_1 - I_E$. Thus, $(\lambda_1 - I_E)(M) \subseteq M$. Next, we consider the map $\lambda_2 : E \to E$ given by $\lambda_2(x_1, x_2, x_3) = (x_1 + \eta(x_2), x_2, x_3)$. This map λ_2 is also an automorphism of E. Thus, as explained before, M is invariant under $\lambda_2 - I_E$, too; that is, $(\lambda_2 - I_E)(M) \subseteq M$.

Let $x = (x_1, x_2, x_3) \in M$. Then $(\lambda_1 - I_E)(x) = (0, \sigma(x_1), 0) \in M$. Similarly, we have $(\lambda_2 - I_E)(x) = (\eta(x_2), 0, 0) \in M$. This gives $(\lambda_1 - I_E)(\eta(x_2), 0, 0) = (0, \sigma\eta(x_2), 0) = (0, x_2, 0) \in M$. Thus, $\pi_2(M) \subseteq M$. Similarly, $(\lambda_2 - I_E)(0, \sigma(x_1), 0) = (\eta\sigma(x_1), 0, 0) = (x_1, 0, 0) \in M$. Therefore, $\pi_1(M) \subseteq M$. This yields that $(0, 0, x_3) \in M$; that is, $\pi_3(M) \subseteq M$. This shows that

$\pi_1(M) \oplus \pi_2(M) \oplus \pi_3(M) \subseteq M$, and therefore, $M = \pi_1(M) \oplus \pi_2(M) \oplus \pi_3(M)$. Hence, $M = (M \cap E_1) \oplus (M \cap E_2) \oplus (M \cap E_3)$. $\qquad\square$

As a consequence of the preceding decomposition, we have the following for socle of an indecomposable automorphism-invariant module.

Corollary 4.42 *If M is an indecomposable automorphism-invariant right module over any ring R, then $\operatorname{Soc}(M)$ is square-free.*

Proof Let M be an indecomposable automorphism-invariant module. Suppose M has two isomorphic simple submodules S_1 and S_2. Then $E(M) = E_1 \oplus E_2 \oplus E_3$, where $E_1 = E(S_1), E_2 = E(S_2)$ and $E_1 \cong E_2$. By Lemma 4.41, M decomposes as $M = (M \cap E_1) \oplus (M \cap E_2) \oplus (M \cap E_3)$, a contradiction to our assumption that M is indecomposable. Hence, $\operatorname{Soc}(M)$ is square-free. $\qquad\square$

Next, we have the following for any indecomposable semi-Artinian automorphism-invariant module.

Corollary 4.43 *Let R be any ring, and let M be any indecomposable semi-Artinian automorphism-invariant right R-module. Then one of the following statements holds:*

1. *M is uniform and quasi-injective.*
2. *Any simple submodule S of M has identity as its only automorphism.*

Proof Let M be an indecomposable semi-Artinian automorphism-invariant right R-module. Since M is semi-Artinian, $\operatorname{Soc}(M) \neq 0$. By Corollary 4.42, we know that $\operatorname{Soc}(M)$ is square-free. Suppose S is a simple submodule of M. Now $D = \operatorname{End}(S)$ is a division ring.

Suppose $|D| > 2$. Then there exists a $\sigma \in D$ such that $\sigma \neq 0$ and $\sigma \neq I_S$. Then $I_S - \sigma$ is an automorphism of S. Let $E = E(M)$ and $E_1 = E(S) \subseteq E$. Then $E = E_1 \oplus E_2$ for some submodule E_2 of E. Let $\sigma_1 \in \operatorname{End}(E_1)$ be an extension of σ. Then σ_1 is an automorphism of E_1 and $(I_{E_1} - \sigma_1)(S) = (I_S - \sigma)(S) \neq 0$. Hence, $I_{E_1} - \sigma_1$ is an automorphism of E_1. Thus, by Lemma 4.39, $M = (M \cap E_1) \oplus (M \cap E_2)$. As M is indecomposable, we must have $M = M \cap E_1$. Therefore, M is uniform. Then $\operatorname{End}(E(M))$ is a local ring. Therefore, for any $\alpha \in \operatorname{End}(E(M))$, α is an automorphism or $I - \alpha$ is an automorphism. In any case, $\alpha(M) \subseteq M$. Therefore, M is quasi-injective.

Now, if M is not uniform, then $|D| = 2$; that is, $D = \operatorname{End}(S) \cong \mathbb{Z}/(2\mathbb{Z})$. In this case, the only automorphism of S is the identity automorphism. $\qquad\square$

Definition 4.44 A ring R is said to be of right automorphism-invariant type (in short, RAI-type), if every finitely generated indecomposable right R-module is automorphism-invariant.

We proceed to give the structure of right Artinian rings of RAI-type.

Lemma 4.45 *Let R be a right Artinian ring of RAI-type. Let $e \in R$ be an indecomposable idempotent such that eR is not uniform. Let A be a right ideal contained in $\mathrm{Soc}(eR)$. Then $\mathrm{Soc}(eR) = A \oplus A'$, where A' has no simple submodule isomorphic to a simple submodule of A and eR/A' is quasi-projective.*

Proof As $\mathrm{Soc}(eR)$ is square-free, $\mathrm{Soc}(eR) = A \oplus A'$, where A' has no simple submodule isomorphic to a simple submodule of A. If for some $ere \in eRe$, $ereA' \not\subseteq A'$, then for some minimal right ideal $S \subset A'$, $ereS \not\subseteq A'$. This gives that S is isomorphic to a simple submodule contained in A, a contradiction. Hence, eR/A' is quasi-projective.   □

Lemma 4.46 *Let R be a right Artinian ring of RAI-type. Then any uniserial right R-module is quasi-projective.*

Proof Let A be a uniserial right R-module with composition length $l(A) = n \geq 2$. We will prove the result by induction. Suppose first that $n = 2$. In this case, we can take $J(R)^2 = 0$. For some indecomposable idempotent $e \in R$, we have $A \cong eR/B$ for some $B \subseteq \mathrm{Soc}(eR)$. By Lemma 4.45, A is quasi-projective.

Now consider $n > 2$, and assume that the result holds for $n - 1$. Let $0 \neq \sigma : A \to A/C$ be a homomorphism where $C \neq 0$. Suppose σ cannot be lifted to a homomorphism $\eta : A \to A$. Let $F = \mathrm{Soc}(A)$. Then $F \subseteq \mathrm{Ker}(\sigma)$. We get a mapping $\bar{\sigma} : A/F \to A/C$. By the induction hypothesis, there exists a homomorphism $\bar{\eta} : A/F \to A/F$ such that $\bar{\sigma} = \pi\bar{\eta}$, where $\pi : A/F \to A/C$ is a natural homomorphism.

Let $M = A \times A$, and $N = \{(a,b) \in M : \bar{\eta}(a + F) = b + F\}$. Then N is a submodule of M. Now there exist elements $x \in A$ and indecomposable idempotent $e \in R$ such that $A = xR$ and $xe = x$. Fix an element $y \in A$ such that $\bar{\eta}(x + F) = y + F$ and $ye = y$. Set $z = (x, y)$. Then $z \in N$ and $N_1 = zR$ is local. Let π_1, π_2 be the associated projections of M onto the first and second components of M, respectively. Then $\pi_1(N_1) = A$.

Now, we claim that N_1 is uniserial. If N_1 is not uniform, then $\mathrm{Soc}(N_1) = \mathrm{Soc}(M)$. Therefore, $\mathrm{Soc}(N_1)$ is not square-free, which is a contradiction by Lemma 4.42. Thus, N_1 is uniform. It follows that N_1 embeds in A under π_1 or π_2. Hence, N_1 is uniserial. As $\pi_1(N_1) = A$, and $l(N_1) \leq l(A)$, it follows that $\pi_1|_{N_1}$ is an isomorphism. Thus, given any $x \in A$, there exists a unique $y \in A$ such that $(x, y) \in N_1$. We get a homomorphism $\lambda : A \to A$ such that

$\lambda(x) = y$ if and only if $(x, y) \in N_1$. Clearly, λ lifts $\overline{\eta}$, and hence, it also lifts σ. This proves that A is quasi-projective. $\qquad\square$

Lemma 4.47 *Let R be a right Artinian ring of RAI-type. Let A_R be any uniserial module. Then the rings of endomorphisms of different composition factors of A are isomorphic.*

Proof Let A be a uniserial right R-module with $l(A) = 2$. Let $C = ann_r(A)$ and $\overline{R} = R/C$. As A_R is quasi-projective, A is a projective $\overline{R}$-module. Thus, there exists an indecomposable idempotent $e \in R$ such that $A \cong \overline{e}\overline{R}$. As $\overline{R}$ embeds in a finite direct sum of copies of A, there exists an indecomposable idempotent $f \in R$ such that $\mathrm{Soc}(A) \cong (\overline{f}\overline{R})/(\overline{fJ(R)})$, $\overline{e}\overline{J(R)} = \overline{exf}\overline{R}$ for some $x \in J(R)$. We get an embedding $\sigma: (eRe)/(eJ(R)e) \to (fRf)/(fJ(R)f)$ defined as $\sigma(\overline{ere + eJ(R)e}) = \overline{fr'f + fJ(R)f}$ whenever $\overline{ere}\overline{exf} = \overline{exf}\overline{fr'f}$; $ere \in eRe$, $fr'f \in fRf$. Let $z = fvf \in fRf$. We get an $\overline{R}$-homomorphism $\eta: \overline{eJ(R)} \to \overline{eJ(R)}$ such that $\eta(\overline{exf}) = \overline{exf}\overline{fvf}$. As $\overline{e}\overline{R}$ is quasi-injective, there exists an $\overline{R}$-homomorphism $\lambda: \overline{e}\overline{R} \to \overline{e}\overline{R}$ extending η. Now $\lambda(\overline{e}) = \overline{ere}$ for some $r \in R$. Then $\overline{ere}\overline{exf} = \lambda(\overline{exf}) = \eta(\overline{exf}) = \overline{exf}\overline{fvf}$, which gives that σ is onto. Hence, $(eRe)/(eJ(R)e) \cong (fRf)/(fJ(R)f)$. Thus, the result holds whenever $l(A)=2$. If $l(A)= n > 2$, the result follows by induction on n. $\qquad\square$

Lemma 4.48 *Let R be a right Artinian ring of RAI-type. Then we have the following.*

1. *Let D be a division ring and $x \in R$. Let xR be a local module such that for any simple submodule S of $\mathrm{Soc}(xR)$, $D = \mathrm{End}(S)$. Then $\mathrm{End}(xR/xJ(R)) \cong D$.*
2. *Let xR be a local module and $D = \mathrm{End}(xR/xJ(R))$, where $x \in R$. Then $\mathrm{End}(S) \cong D$ for every composition factor S of xR.*
3. *Let xR, yR be two local modules where $x, y \in R$. If $\mathrm{End}(xR/xJ(R)) \not\cong \mathrm{End}(yR/yJ(R))$, then $\mathrm{Hom}(xR, yR) = 0$.*

Proof 1. There exists an $n \geq 1$ such that $xJ(R)^n = 0$ but $xJ(R)^{n-1} \neq 0$. If $n = 1$, then xR is simple, so the result holds. We apply induction on n. Suppose $n > 1$, and assume that the result holds for $n - 1$. Now $xJ(R)J(R)^{n-1} = 0$, but $xJ(R)J(R)^{n-2} \neq 0$. Therefore, there exists an element $y \in xJ(R)$ such that yR is local and $yJ(R)^{n-1} = 0$ but $yJ(R)^{n-2} \neq 0$. By the induction hypothesis, $\mathrm{End}(yR/yJ(R)) \cong D$. In fact, for any simple submodule S' of $eJ(R)/xJ(R)^2$, $\mathrm{End}(S') \cong D$.

Consider the local module $M = xR/xJ(R)^2$. Let S' be a simple submodule of M. Then $\mathrm{Soc}(M) = S' \oplus B$ for some $B \subset \mathrm{Soc}(M)$. Then $\mathrm{End}(S') \cong D$. As $A = M/B$ is uniserial, $\mathrm{Soc}(A) \cong S'$ and $A/AJ(R) \cong xR/xJ(R)$. By Lemma 4.47, $\mathrm{End}(xR/xJ(R)) \cong D$.

2. Let S be a simple submodule of $\mathrm{Soc}(xR)$, and let B be a complement of S in xR. Then $\overline{xR} = xR/B$ is uniform and $\mathrm{Soc}(\overline{xR}) \cong S$. By (1), $\mathrm{End}(S) \cong \mathrm{End}\,(\overline{xR})/(\overline{xJ(R)}) \cong \mathrm{End}(xR/xJ(R)) = D$. Hence, $\mathrm{End}(S) \cong D$ for any simple submodule S of xR. Let S_1 be any composition factor of xR. Then there exists a local submodule yR of xR such that $S_1 \cong yR/yJ(R)$. By (1), $\mathrm{End}(S_1) \cong \mathrm{End}(S) \cong D$, where S is a simple submodule of yR.

3. It is immediate from (2).   □

Now, we are ready to give the structure of indecomposable right Artinian rings of RAI-type.

Theorem 4.49 *Let R be an indecomposable right Artinian ring of RAI-type. Then the following hold.*

1. *There exists a division ring D such that $\mathrm{End}(S) \cong D$ for any simple right R-module S. In particular, $R/J(R)$ is a direct sum of matrix rings over D.*
2. *If $D \not\cong \mathbb{Z}/2\mathbb{Z}$, then every finitely generated indecomposable right R-module is quasi-injective. In this case, R is right serial.*

Proof 1. Let $e \in R$ be an indecomposable idempotent and $D = eRe/eJ(R)e$. By preceding lemma, every composition factor S of eR satisfies $\mathrm{End}(S) \cong D$. Now $R_R = \oplus_{i=1}^{n} e_i R$, where e_i are orthogonal indecomposable idempotents with $e_1 = e$. Let A be the direct sum of those $e_j R$ for which $(e_j Re_j)/(e_j J(R)e_j) \cong D$. Consider any e_k for which $(e_k Re_k)/(e_k J(R)e_k) \not\cong D$. It follows from Lemma 4.48(iii) that $Ae_k R = 0 = e_k RA$. Consequently, $A = e_k R$, and we get that $R = A \oplus B$ for some ideal B. As R is indecomposable, we get $R = A$. This proves (1).

2. Suppose $D \not\cong \mathbb{Z}/2\mathbb{Z}$. It follows from Corollary 4.43 that every indecomposable right R-module is uniform and quasi-injective. In particular, if $e \in R$ is an indecomposable idempotent, then any homomorphic image of eR is uniform, which gives that eR is uniserial. Hence, R is right serial.   □

Theorem 4.50 *[92] Let R be a right Artinian ring such that $J(R)^2 = 0$. If every finitely generated indecomposable right R-module is local, then R satisfies the following conditions.*

1. *Every uniform right R-module is either simple or is injective with composition length 2.*
2. *R is a left serial ring.*
3. *For any indecomposable idempotent $e \in R$, either $eJ(R)$ is homogeneous or $l(eJ(R)) \leq 2$.*

Conversely, if R satisfies (1), (2), (3) and $l(eJ(R)) \leq 2$, then every finitely generated indecomposable right R-module is local.

Example 4.51 Let $R = \begin{bmatrix} \mathbb{F} & \mathbb{F} & \mathbb{F} \\ 0 & \mathbb{F} & 0 \\ 0 & 0 & \mathbb{F} \end{bmatrix}$, where $\mathbb{F} = \mathbb{Z}/2\mathbb{Z}$.

Then R is a left serial ring. We have already seen that $e_{11}R$ is an indecomposable module that is automorphism-invariant but not quasi-injective. It follows from Theorem 4.50 that every finitely generated indecomposable right R-module is local. Thus, the only indecomposable modules that are not simple are the homomorphic images of $e_{11}R$, which are $e_{11}R$, $(e_{11}R)/(e_{12}\mathbb{F})$ and $(e_{11}R)/(e_{13}\mathbb{F})$. These are all automorphism-invariant. It follows from Theorem 4.50 that any finitely generated indecomposable right R-module is local. Thus, this ring R is an example of a ring where every finitely generated indecomposable right R-module is automorphism-invariant.

Example 4.52 Let $\mathbb{F} = \mathbb{Z}/2\mathbb{Z}$ and $R = \begin{bmatrix} \mathbb{F} & \mathbb{F} & \mathbb{F} & \mathbb{F} \\ 0 & \mathbb{F} & 0 & 0 \\ 0 & 0 & \mathbb{F} & 0 \\ 0 & 0 & 0 & \mathbb{F} \end{bmatrix}$.

This ring R is left serial and $J(R)^2 = 0$. Now $e_{11}J(R) = e_{12}\mathbb{F} \oplus e_{13}\mathbb{F} \oplus e_{14}\mathbb{F}$, a direct sum of non-isomorphic minimal right ideals. It follows from condition (3) in Theorem 4.50 that there exists a finitely generated indecomposable right R-module that is not local. We have $E_1 = E(e_{12}\mathbb{F})$, $E_2 = E(e_{13}\mathbb{F})$, $E_3 = E(e_{14}\mathbb{F})$; each of them has composition length 2. Now $e_{11}R$ has two homomorphic images $A_1 = (e_{11}R)/(e_{14}\mathbb{F})$ and $A_2 = (e_{11}R)/(e_{12}\mathbb{F})$ such that $\text{Soc}(A_1) \cong e_{12}\mathbb{F} \oplus e_{13}\mathbb{F}$ and $\text{Soc}(A_2) \cong e_{13}\mathbb{F} \oplus e_{14}\mathbb{F}$. So we get $B_1 \subseteq E_1 \oplus E_2 \subseteq E_1 \oplus E_2 \oplus E_3$ such that $A_1 \cong B_1$. Similarly, we have $A_2 \cong B2 \subseteq E_2 \oplus E3$. Let $E = E_1 \oplus E_2 \oplus E_3$. Its only automorphism is I_E. Thus, any essential submodule of E is automorphism-invariant. Now $B = B_1 + B_2 \subseteq_e E$, so B is automorphism-invariant and B is not local. We prove that B is indecomposable. We have $B_1 \cap B_2 = e_{13}\mathbb{F}$. Notice that any submodule of $E_1 \oplus E_2$ that is indecomposable and not uniserial is B_1. Suppose a simple submodule S of B is a summand of B. But $S \subset B_1$ or $S \subset B_2$; therefore, B_1

or B_2 decomposes, which is a contradiction. As $l(B) = 5$, B has a summand C_1 with $l(C_1) = 2$. Then with C_1 being uniserial, it equals one of E_i.

Without loss of generality, assume $C_1 = E_1$. Then $B = C_1 \oplus C_2$, where $\text{Soc}(C_2) \cong B_2$. As C_2 has no uniserial submodule of length 2, the projection of B_1 in C_2 equals $\text{Soc}(C_2)$, we get B_1 is semisimple, which is a contradiction.

Similarly, other cases follow. Hence, B is indecomposable.

4.4 Rings Whose Cyclic Modules Are Automorphism-Invariant

Characterizing rings via homological properties of their cyclic modules is a problem that has been studied extensively in the last 50 years. The most recent account of results related to this prototypical problem may be found in [67].

Theorem 4.53 *Let R be a ring over which every cyclic right R-module is automorphism-invariant. Then $R \cong S \times T$, where S is a semisimple Artinian ring, and T is a right square-free ring such that, for any two closed right ideals X and Y of T with $X \cap Y = 0$, $\text{Hom}(X, Y) = 0$. In particular, all idempotents of T are central.*

Proof By the proof of Theorem 3.6, we have a decomposition $R_R = A \oplus B \oplus B' \oplus C$, where $A \cong B$, B' is isomorphic to a submodule of B, and C is square-free and $A \oplus B \oplus B'$ and C are orthogonal. Let Z be a right ideal in A. Then $R/Z \cong A/Z \oplus B \oplus B' \oplus C$ is automorphism-invariant by assumption. Then A/Z is B-injective, whence A-injective. Similarly, all factors of B, B', and C are A-injective as well.

Now, A is a cyclic projective module all of whose factors are A-injective (and, in particular, quasi-injective). So, by [24, corollary 9.3 (ii)], $A = U_1 \oplus \cdots \oplus U_n$, where U_i are uniform modules. Take an arbitrary nonzero cyclic submodule U of U_i, for any i. Since U is a sum of factors of A, B, B' and C, it contains a nonzero factor of one of them, call it U'. By the preceding paragraph, U' is A-injective, so it splits in U_i. Thus, $U' = U = U_i$, showing that U_i is simple, whence $A \oplus B \oplus B'$ is semisimple. Since $A \oplus B \oplus B'$ and C are orthogonal projective modules and the former is now semisimple, there are no nonzero homomorphisms between them. Therefore, $A \oplus B \oplus B'$ and C are ideals. So now we have the ring direct sum $R = S \oplus T$, where $S = A \oplus B \oplus B'$ and $T = C$.

Now let X and Y be closed right ideals of T such that $X \cap Y = 0$, and let $f: X \to Y$ be any homomorphism. Set $Y' = f(X)$. This induces an isomorphism $\overline{f}: X/K \to Y'$, where $K = \text{Ker}(f)$. It is clear that X/K is a closed submodule of T/K. Also, since T_T is square-free, K is essential in X. Choose a complement U/K of $X/K \oplus (Y' \oplus K)/K$ in T/K. Since T/K is automorphism-invariant by assumption and $X/K \cong Y' \cong (Y' \oplus K)/K$ by the last part of Lemma 4.5, $(Y' \oplus K)/K$ is closed in T/K. Applying Lemma 4.45, we obtain $T/K = X/K \oplus (Y' \oplus K)/K \oplus U/K$. Since $Y' \cap (X + U) \subseteq Y' \cap K = 0$, we have $T = Y' \oplus (X + U)$. So Y'_T is projective, whence the map f splits. However, since K is essential in X, we have $f = 0$. So Hom $(X, Y) = 0$. In particular, if $T_T = X \oplus Y$, we have $XY = YX = 0$, whence X and Y are ideals. $\qquad\qquad\qquad\qquad\qquad\qquad\qquad\qquad\qquad\qquad\qquad\qquad$ □

Using an alternative argument to the one in the second paragraph of the preceding proof, we can generalize the decomposition in the theorem as follows:

Proposition 4.54 *Let M be a module satisfying either of the following conditions:*

1. *M is cyclic with all factors automorphism-invariant and generates its cyclic subfactors, or*
2. *M is an automorphism-invariant module whose every 2-generated subfactor is automorphism-invariant.*

Then $M = X \oplus Y$, where X is semisimple, Y is square-free and X and Y are orthogonal.

Proof First, note that, by the proof of Theorem 3.6, we have a decomposition $M = A \oplus B \oplus B' \oplus C$, where $A \cong B$, B' embeds in B and C is square-free and orthogonal to $A \oplus B \oplus B'$.

1. In this case, in the same way as in the first paragraph of the proof of Theorem 4.53, all factors of the modules B ($\cong A$), B' and C are A-injective. Now let A' be any factor of A and D be a cyclic submodule of A'. Since D is generated by M, $D = D_1 + \cdots + D_n$, where each D_i is a factor of B, B' or C. Since D_1 is A-injective (whence A'-injective), $D_1 \oplus D'_1 = A'$ for some submodule D'_1 of A'. Letting $\pi : D_1 \oplus D'_1 \to D'_1$ be the obvious projection, we have $D = D_1 \oplus (\pi(D_2) + \cdots + \pi(D_n))$. Each $\pi(D_k)$ again being a factor of B, B' of C, it is A-injective, hence D'_1-injective. By induction on n, we obtain that D is a direct sum of A-injective cyclic modules. Then D is A-injective. Now we have shown that each cyclic subfactor of A is A-injective. By [24, corollary 7.14], A is semisimple. Therefore, $A \oplus B \oplus B'$ is semisimple as well. Now set $X = A \oplus B \oplus B'$ and $Y = C$.

2. Let $D \subseteq L$ be submodules of A with L/D cyclic, and let T be a cyclic submodule of B. By assumption, $L/D \oplus T$ is automorphism-invariant,

whence L/D is T-injective. Then cyclic subfactors of A are B-injective, hence A-injective. Again, by [24, corollary 7.14], A is semisimple. The conclusion follows in the same way as before. $\qquad\square$

4.5 Rings Whose Each One-Sided Ideal Is Automorphism-Invariant

Definition 4.55 A ring R is called a right a-ring if each right ideal of R is automorphism-invariant.

Rings whose each right ideal is quasi-injective are known as right q-rings. Clearly, any right q-ring is a right a-ring. Recall that right q-rings are precisely those right self-injective rings for which every essential right ideal is a two-sided ideal [67]. So, in particular, any commutative self-injective ring is a q-ring and, hence, an a-ring. Now we would like to present some examples of right a-rings that are not right q-rings. First, we have the following useful observation.

Lemma 4.56 *A commutative ring is an a-ring if and only if it is an automorphism-invariant ring.*

Proof Let R be a commutative automorphism-invariant ring, and let I be an ideal of R. There exists an ideal U of R such that $I \oplus U$ is essential in R. Then $E(R) = E(I \oplus U)$. Let φ be an automorphism of $E(R)$. Clearly, $\varphi(1) \in R$. Now, for all $x \in I \oplus U$, we have $\varphi(x) = \varphi(1)x \in I \oplus U$. So $\varphi(I \oplus U) \leq I \oplus U$, which implies that $I \oplus U$ is an automorphism-invariant module. Since direct summand of an automorphism-invariant module is automorphism-invariant, it follows that I is automorphism-invariant. This shows that R is an a-ring. The converse is obvious. $\qquad\square$

In view of the preceding, we have the following example of an a-ring that is not a q-ring.

Example 4.57 Consider the ring R consisting of all eventually constant sequences of elements from $\mathbb{F}_2$ as in Example 2.12. Clearly, R is a commutative automorphism-invariant ring as the only automorphism of its injective envelope is the identity automorphism. Hence, R is an a-ring by the preceding lemma. But R is not a q-ring because R is not self-injective.

Now we will prove some characterizations for right a-rings. These equivalent characterizations will be more convenient to use.

Proposition 4.58 *The following conditions are equivalent for a ring R:*

1. *R is a right a-ring.*
2. *Every essential right ideal of R is automorphism-invariant.*
3. *R is right automorphism-invariant and every essential right ideal of R is a left T-module, where T is a subring of R generated by its unit elements.*

Proof (1) $\Rightarrow$ (2). This is obvious.

(2) $\Rightarrow$ (3). By the hypothesis, R is a right automorphism-invariant ring. Let I be an essential right ideal of R. Then $E(I) = E(R)$. Let T be a subring of R generated by its units. Then T is a subring of $\mathrm{End}(E(R))$, and so $TI = I$.

(3) $\Rightarrow$ (2). Let I be an essential right ideal of R. Then $E(I) = E(R)$. Let φ be an automorphism of $E(R)$. As R is right automorphism-invariant, we have $\varphi(R) = R$, which implies that $\varphi(1)$ is a unit of R. By the assumption, we have $\varphi(1)I \leq I$, and so $\varphi(I) \leq I$. This shows that each essential right ideal of R is automorphism-invariant.

(2) $\Rightarrow$ (1). Let A be any right of R. Let K be a complement of A; then $A \oplus K$ is an essential right ideal of R. By assumption, $A \oplus K$ is automorphism-invariant. Since every direct summand of an automorphism-invariant module is automorphism-invariant, it follows that A is automorphism-invariant. This proves that R is a right a-ring. $\qquad\square$

Corollary 4.59 *Let $R = S \times T$ be a product of rings. Then R is a right a-ring if and only if S and T are right a-rings.*

Lemma 4.60 *Let R be a right a-ring, and let A be a right ideal of R. If there exists a right ideal B of R with $A \cap B = 0$ and $A \cong B$, then:*

1. *A is semisimple and injective.*
2. *A is nonsingular.*

Proof (1) Let A and B be right ideals of a right a-ring R with $A \cap B = 0$ and $A \cong B$. Let D be a complement of $A \oplus B$ in R_R. Then $(A \oplus B) \oplus D \leq^e R_R$. It follows that $E((A \oplus B) \oplus D) \leq^e E(R_R)$. On the other hand, $E((A \oplus B) \oplus D)$ is a direct summand of $E(R_R)$, and so $E((A \oplus B) \oplus D) = E(R_R)$. We have $E((A \oplus B) \oplus D) = E(A) \oplus E(B) \oplus E(D)$. Thus, $E(R_R) = E(A) \oplus E(B) \oplus E(D)$, which means that we have a decomposition $E(R_R) = E(A) \oplus E(B) \oplus C$ for some $C \leq E(R_R)$. Note that $E(A) \cong E(B)$ and R is right automorphism-invariant. By Lemma 4.41, we get

$$R_R = (R \cap E(A)) \oplus (R \cap E(B)) \oplus (R \cap C).$$

We also have $B \cap (R \cap E(A)) = 0$ and $A \cap [(R \cap E(B)) \oplus (R \cap C)] = 0$. Since R is a right a-ring, the modules $B \oplus [R \cap E(A)]$ and $A \oplus [(R \cap E(B)) \oplus (R \cap C)]$ are automorphism-invariant. By Corollary 4.6, B is $[R \cap E(A)]$-injective, and A is $[(R \cap E(B)) \oplus (R \cap C)]$-injective. Note that $A \cong B$. Thus, A is R-injective; that is, A is injective. Let $\varphi \colon A \to B$ be an isomorphism, and let U be a submodule of A. Clearly, $U \cong \varphi(U)$. Let $V = \varphi(U)$. Then $U \cap V = 0$ and $U \cong V$. By a similar argument as before, we have that U is an injective module, and thus, it follows that U is a direct summand of A. This proves that A is semisimple.

(2) Let a be an arbitrary element of $Z(A)$. Then aR is an injective module since it is a direct summand of A. It follows that $aR = eR$ for some $e^2 = e \in R$. Therefore, $e \in Z(A)$ and so $e = 0$. Thus, $a = 0$, which shows $Z(A) = 0$. $\qquad\qquad\qquad\qquad\qquad\qquad\qquad\qquad\qquad\qquad\qquad\qquad$ □

As a consequence of the preceding lemma, we are now ready to prove a useful decomposition theorem for any right a-ring.

Theorem 4.61 *A right a-ring is a direct sum of a square-full semisimple Artinian ring and a right square-free ring.*

Proof We have a decomposition $R_R = A \oplus B \oplus C$ where $A \cong B$ and the module C is square-free, which is orthogonal to $A \oplus B$. Let $X := A \oplus B$ and $Y := C$. Now we proceed to show that X is square-full. Let U be a nonzero arbitrary submodule of X. There exist either nonzero submodules U_1 of U and V_1 of A such that $U_1 \cong V_1$ or nonzero submodules U_2 of U and V_2 of B such that $U_2 \cong V_2$. It follows that U_1^2 or U_2^2 can be embedded in X. This means U contains a square root in X, and hence, X is square-full.

By Lemma 4.60, A and B are injective semisimple modules, and so X is injective and semisimple. Next, we show that X and Y are ideals of R. Let f be a nonzero homomorphism from X to Y. As X is semisimple, $\mathrm{Ker}(f)$ is a direct summand of X. So there exists a submodule L of X such that $X = \mathrm{Ker}(f) \oplus L$. As $\mathrm{Ker}(f) \cap L = 0$, we have $L \cong f(L) \subseteq Y$, a contradiction to the fact that X is orthogonal to Y. Hence, we have $\mathrm{Hom}(X, Y) = 0$. Assume that $\varphi \colon Y \to X$ is a nonzero homomorphism. Then $Y/(\mathrm{Ker}(\varphi)) \cong \mathrm{Im}(\varphi)$ is projective (since $\mathrm{Im}(\varphi)$ is a direct summand of X). It follows that there exists a nonzero submodule K of Y such that $\mathrm{Ker}(\varphi) \cap K = 0$. So $K \cong \varphi(K)$, a contradiction to the orthogonality of X and Y. Therefore, $\mathrm{Hom}(Y, X) = 0$.

Thus, $R = X \oplus Y$, where X is a square-full semisimple Artinian ring and Y is a right square-free ring. $\qquad\qquad\qquad\qquad\qquad\qquad\qquad\qquad\qquad\qquad$ □

As a consequence of the preceding, we have the following.

Corollary 4.62 *An indecomposable ring R containing a square is a right a-ring if and only if R is simple Artinian.*

We denote the ring of $n \times n$ matrices over a ring R by $\mathbb{M}_n(R)$. In the next theorem, we study when matrix rings are right a-rings.

Theorem 4.63 *Let $n > 1$ be an integer. The following conditions are equivalent for a ring R:*

1. $\mathbb{M}_n(R)$ *is a right q-ring.*
2. $\mathbb{M}_n(R)$ *is a right a-ring.*
3. *R is semisimple Artinian.*

Proof The implication $(1) \Rightarrow (2)$ is obvious.

$(2) \Rightarrow (3)$. Let $\mathbb{M}_n(R)$ be a right a-ring. Assume that R is not semisimple Artinian. Then there exists an essential right ideal, say B, of R such that $B \neq R$. Define $E := \left\{ \sum a_{ij}e_{ij} : a_{1j} \in B, 1 \leq j \leq n \text{ and } a_{ij} \in R, 1 \leq i, j \leq n \right\}$ where e_{ij} $(1 \leq i, j \leq n)$ are the units of $\mathbb{M}_n(R)$. Then, clearly, E is an essential right ideal of $\mathbb{M}_n(R)$. Consider the unit $\begin{pmatrix} 0 & 0 & 0 & \cdots & 0 & 1 \\ 0 & 0 & 0 & \cdots & 1 & 0 \\ 0 & 0 & 0 & \cdots & 0 & 0 \\ \vdots & \vdots & \vdots & \ddots & \vdots & \vdots \\ 1 & 0 & 0 & \cdots & 0 & 0 \end{pmatrix}$ of $\mathbb{M}_n(R)$.

Then

$$\begin{pmatrix} 0 & 0 & 0 & \cdots & 0 & 1 \\ 0 & 0 & 0 & \cdots & 1 & 0 \\ 0 & 0 & 0 & \cdots & 0 & 0 \\ \vdots & \vdots & \vdots & \ddots & \vdots & \vdots \\ 1 & 0 & 0 & \cdots & 0 & 0 \end{pmatrix} \begin{pmatrix} 0 & 0 & 0 & \cdots & 0 & 0 \\ 0 & 0 & 0 & \cdots & 0 & 0 \\ 0 & 0 & 0 & \cdots & 0 & 0 \\ \vdots & \vdots & \vdots & \ddots & \vdots & \vdots \\ 0 & 0 & 0 & \cdots & 0 & 1 \end{pmatrix}$$

$$= \begin{pmatrix} 0 & 0 & 0 & \cdots & 0 & 1 \\ 0 & 0 & 0 & \cdots & 0 & 0 \\ 0 & 0 & 0 & \cdots & 0 & 0 \\ \vdots & \vdots & \vdots & \ddots & \vdots & \vdots \\ 0 & 0 & 0 & \cdots & 0 & 0 \end{pmatrix} \notin E.$$

This yields a contradiction (see Proposition 4.58). Hence, R is semisimple Artinian.

$(3) \Rightarrow (1)$. This is obvious. $\qquad\square$

The following example shows that there exists automorphism-invariant rings which are not right a-rings.

Example 4.64 Let $R = \mathbb{Z}_{p^n}$, where p is a prime and $n > 1$. It is well known that R is self-injective. By [129, theorem 8.3], $\mathbb{M}_m(R)$ is right self-injective for all $m > 1$. Thus, for instance, $\mathbb{M}_m(\mathbb{Z}_{p^2})$ is a right automorphism-invariant ring. But $\mathbb{M}_m(\mathbb{Z}_{p^2})$ is not a right a-ring for any $m > 1$ in the view of preceding theorem as $\mathbb{Z}_{p^2}$ is not semisimple Artinian. This example also shows that being a right a-ring is not a Morita invariant property.

Now we will consider some special classes of rings – for example, simple, semiprime, prime and CS and characterize as to when these rings are right a-rings.

Lemma 4.65 *Let A and B be right ideals of a right a-ring R with $A \cap B = 0$. Then the following conditions hold:*

1. *If $\varphi: A \to B$ is a nonzero homomorphism, then*

    (i) *$\varphi(A)$ is a semisimple module.*
    (ii) *$\varphi(A)$ is simple if B is uniform.*

2. *If e is a nontrivial idempotent of R such that $eR(1 - e) \neq 0$, then $\mathrm{Soc}(eR) \neq 0$.*

Proof 1(i). Let U be an arbitrary essential submodule of B. Then $E(U) = E(B)$ and $U \oplus A$ is automorphism-invariant. It follows that U is A-injective. On the other hand, there exists a homomorphism $\bar{\alpha}: E(A) \to E(B)$ such that $\bar{\alpha}|_A = \varphi$. It follows that $\bar{\alpha}(A) \leq U$ and so $\varphi(A) \leq U$. This shows that $\varphi(A) \leq \mathrm{Soc}(B)$.

(ii) If B is uniform, then from (i), it follows easily that $\varphi(A)$ is simple.

2. Assume that $eR(1-e) \neq 0$. There exists $r_0 \in R$ such that $er_0(1-e) \neq 0$. Consider the homomorphism $\beta: (1 - e)R \to eR$ defined by $\beta((1 - e)x) = er_0(1 - e)x$. Clearly, β is well defined and $\mathrm{Im}(\beta) \neq 0$. By 1(i), we have $\mathrm{Im}(\beta) \leq \mathrm{Soc}(eR)$. Hence $\mathrm{Soc}(eR) \neq 0$. $\square$

Theorem 4.66 *A right a-ring is von Neumann regular if and only if it is semiprime.*

Proof Let R be a right a-ring. If R is von Neumann regular, then it is well known that R is semiprime. Conversely, assume that R is semiprime. As R is a right a-ring, in particular, R_R is automorphism-invariant. By Theorem 3.3, $R/J(R)$ is von Neumann regular and $J(R) = Z(R_R)$. Now we proceed to show that $J(R) = 0$. In fact, for any $x \in J(R)$, there exists an essential right ideal E of R such that $xE = 0$. Since R is a right a-ring, $uE \leq E$ for all units u in R by Lemma 4.58. It follows that $(RxR)E \leq E$ and so $(xRxR)E \leq xE = 0$, and so either $xRxR \leq P$ or $E \leq P$ for all prime ideal P of R. Let $\{P_i\}_{i \in I}$ and $\{P_j\}_{j \in J}$ be families of all prime ideals of R such that $xRxR \leq P_i$ for all $i \in I$ and $xRxR \nleq P_j$ for all $j \in J$. Taking $X = \cap_{i \in I} P_i$ and $Y = \cap_{j \in J} P_j$. Since R is semiprime, $X \cap Y = 0$. Moreover, we have $E \leq Y$ and so $Y \leq^e R_R$. If $xRxR \neq 0$, there exists $r_1, r_2 \in R$ such that $xr_1xr_2 \neq 0$. Then there exists a $y \in R$ such that $xr_1xr_2y \neq 0$ and $xr_1xr_2y \in Y$, a contradiction. Thus, $xRxR = 0$. Furthermore, as R is semiprime, we have $x = 0$. This completes the proof. $\qquad\square$

Theorem 4.67 *Every right a-ring is stably finite.*

Proof Let R be a right a-ring. Then $R = S \times T$, where S_S is semisimple Artinian and T_T is square-free. Clearly, T is a right quasi-duo ring. Then $\mathbb{M}_n(R) = \mathbb{M}_n(S \oplus T)$. Thus, $\mathbb{M}_n(R) \cong \mathbb{M}_n(S) \oplus \mathbb{M}_n(T)$. Clearly, $\mathbb{M}_n(S)$ is directly finite. Now we proceed to show that $\mathbb{M}_n(T)$ is directly finite. Let $\{M_i\}$ be the set of maximal right ideals of the quasi-duo ring T. Then each M_i is a two-sided ideal and $J(T) = \cap M_i$. Clearly, each T/M_i is a division ring. Thus, $\mathbb{M}_n(T)/\mathbb{M}_n(M_i) \cong \mathbb{M}_n(T/M_i)$ is a simple Artinian ring that is clearly directly finite. Consider the natural ring homomorphism $\varphi \colon \mathbb{M}_n(T) \longrightarrow \prod_i \mathbb{M}_n(T/M_i)$. We have $\mathrm{Ker}(\varphi) = \mathbb{M}_n(J(T)) = J(\mathbb{M}_n(T))$. Since each $\mathbb{M}_n(T/M_i)$ is directly finite, $\prod_i \mathbb{M}_n(T/M_i)$ is directly finite, and consequently, $\mathbb{M}_n(T)/J(\mathbb{M}_n(T))$ is directly finite being a subring of a directly finite ring. Hence, $\mathbb{M}_n(T)$ is directly finite. Thus, $\mathbb{M}_n(R)$ is directly finite, and therefore, R is stably finite. $\qquad\square$

Corollary 4.68 *Every von Neumann regular right a-ring is unit-regular.*

Corollary 4.69 *The ring of linear transformations $R := \mathrm{End}(V_D)$ of a vector space V over a division ring D is a right a-ring if and only if the vector space is finite-dimensional.*

Proof If V is an infinite-dimensional vector space over D then $\mathrm{End}(V_D)$ is not directly finited. So the result follows from Theorem 4.67. $\qquad\square$

Proposition 4.70 *Let R be a semiprime right a-ring with zero socle. Then R is strongly regular.*

Proof Assume that R is a semiprime right a-ring. Clearly, R is von Neumann regular. Let e be an idempotent in R. Suppose $(1 - e)Re \neq 0$. Then

$\text{Soc}((1-e)R) \neq 0$, a contradiction. Hence, $(1-e)Re = 0$, and this shows that e is a central idempotent (see [40, lemma 2.33]). Because every idempotent of R is central, R is strongly regular. $\qquad\square$

Theorem 4.71 *Let R be a prime ring. Then R is a right a-ring if and only if R is a simple Artinian ring.*

Proof Assume that R is a prime right a-ring. In view of Theorem 4.61, we obtain that either R is a simple Artinian ring or R is a square-free ring. So it suffices to consider the case that R is a square-free prime right a-ring. By Theorem 4.66, R is a von Neumann regular ring. Since R is square-free, all idempotents of R are central, and hence, R is a strongly regular ring. Now, as every prime strongly regular ring is a division ring, the result follows. $\qquad\square$

In particular, from the preceding theorem, it follows that every simple right a-ring is Artinian.

A module M is called a *weak CS module* if every semisimple submodule of M is essential in a direct summand of M.

Proposition 4.72 *Let R be a right weak CS right a-ring. If e is a primitive idempotent of R such that $eR(1-e) \neq 0$, then eRe is a division ring and $eR(1-e)$ is the only proper R-submodule of eR.*

Proof By Lemma 4.65, $\text{Soc}(eR) \neq 0$. Since R is right automorphism-invariant, R is right C2 by Theorem 4.4. Now, we know that eR is also a weak CS module. First, we show that $\text{Soc}(eR)$ is a simple module that is essential in eR. Since eR is a weak CS module, $\text{Soc}(eR)$ is essential in a direct summand of eR. But eR is an indecomposable module, which implies that $\text{Soc}(eR)$ is essential in eR. For any nonzero arbitrary element $a \in \text{Soc}(eR)$, we obtain that aR is essential in eR (because eR is an indecomposable weak CS module). It follows that $\text{Soc}(eR) \leq aR$, and so $\text{Soc}(eR) = aR$. Thus, $\text{Soc}(eR)$ is a simple module. Therefore, eR is uniform. Since a uniform automorphism-invariant module is quasi-injective, eR is quasi-injective. Thus, $eRe \cong \text{End}(eR)$ is a local ring; that is, e is a local idempotent of R.

Next we show that $eR(1-e)$ is the only proper submodule of eR. Since $eR(1-e) \neq 0$, one infers $eR(1-e) \subset \text{Soc}(eR)$ by Lemma 4.65. Hence,

$$eR(1-e) = \text{Soc}(eR)(1-e).$$

We next show that $eJ(R)e$ is a submodule of eR. Since R is right automorphism-invariant, $J(R) = Z(R_R)$, and so $J(R)\text{Soc}(eR) = 0$. Now

$(eJ(R)e)\operatorname{Soc}(eR) = eJ(R)\operatorname{Soc}(eR) = 0$, and so $(eJ(R)e)(eR(1-e)) = 0$. On the other hand, we have

$$eJ(R)eR = eJ(R)e(Re + R(1-e)) = eJ(R)eRe \subset eJ(R)e.$$

Hence, $eJ(R)e$ is an R-submodule of eR. Since $\operatorname{Soc}(eR)$ is simple, we have $eJ(R)e \cap \operatorname{Soc}(eR) = 0$ or $\operatorname{Soc}(eR) \le eJ(R)e$. Suppose $\operatorname{Soc}(eR) \le eJ(R)e$. Then $eR(1-e) = \operatorname{Soc}(eR)(1-e) \le eJ(R)e(1-e) = 0$, a contradiction. It follows that $eJ(R)e \cap \operatorname{Soc}(eR) = 0$. Thus, $eJ(R)e = 0$.

Let I be a proper submodule of eR. Since eR is local, $I \le eJ(R)$, and so $Ie = 0$. On the other hand, we have $I(1-e) \le eR(1-e)$, which implies that $I \le eR(1-e) = \operatorname{Soc}(eR)$. Thus, $I = 0$ or $I = \operatorname{Soc}(eR)$. In particular, we have $\operatorname{Soc}(eR)e = 0$. Therefore, $eR(1-e) = \operatorname{Soc}(eR)(1-e) = \operatorname{Soc}(eR)$. □

As a consequence, we have the following.

Theorem 4.73 *Let R be an indecomposable, nonlocal ring. The following conditions are equivalent:*

1. *R is a right q-ring.*
2. *R is a right CS right a-ring.*

Now we would like to describe the structure of right a-rings. For a division ring D, we first set the following notations:

$$H(n; D; \alpha) = \begin{bmatrix} D & V & 0 & & & 0 \\ 0 & D & V & 0 & & 0 \\ & & D & V & 0 & \\ & & & D & V & 0 \\ & & & & D & V \\ V(\alpha) & 0 & & & & D \end{bmatrix}, \text{ where } V \text{ is one-dimensional both}$$

as a left D-space and a right D-space, $V(\alpha)$ is also a one-dimensional left D-space as well as a right D-space with right scalar multiplication twisted by an automorphism α of D, i.e. $vd = v \cdot \alpha(d)$ for all $v \in V$, $d \in D$, and

$$G_n(n; \Delta; P) := \begin{pmatrix} D & V & & & & \\ & D & V & & & \\ & & D & V & & \\ & & & \cdot & \cdot & \cdot \\ & & & & \cdot & \cdot \\ & & & & \cdot & D & V \\ & & & & & & \Delta \end{pmatrix},$$

where V is as before and Δ is a ring with maximal essential right ideal P such that $D = \Delta/P$ is a division ring.

Now, using the preceding defined notations, we give the following description of right a-rings.

Theorem 4.74 *Let $n \geq 1$ be an integer, let $D_1, D_2, \ldots, D_n$ be division rings, and let Δ be a right a-ring with all idempotents central and an essential ideal, say P, such that Δ/P is a division ring and the right Δ-module Δ/P is not embeddable into Δ_Δ. Next, let V_i be a D_i-D_{i+1}-bimodule such that*

$$dim(_{D_i}\{V_i\}) = dim(\{V_i\}_{D_{i+1}}) = 1$$

for all $i = 1, 2, \ldots, n - 1$, and let V_n be a D_n-Δ-bimodule such that $V_n P = 0$ and

$$dim(_{D_n}\{V_n\}) = dim(\{V_n\}_{\Delta/P}) = 1.$$

Then $R := G_n(D_1, \ldots, D_n, \Delta, V_1, \ldots, V_n)$ is a right a-ring.

Proof Let $1 \leq i \leq n + 1$, and let e_i be the matrix whose (i, i)-entry is equal to 1 and all the other entries are 0. It is easy to see that $e_j R e_{j+1}$ are minimal right ideals of R for all $j = 1, 2, \ldots, n$. Let K be a right ideal of the ring Δ, and let $\widehat{K}$ to be the set of all matrices whose $(n + 1, n + 1)$-entries are from K and all the other entries are 0. Take $1 \leq i \leq n$ and a right ideal K of Δ. We are going to adapt the techniques of [10, proposition 2.16]. First, we have the following facts, which will be used throughout in the proof:

Fact 1 $e_i R$ and $\widehat{K}$ are relatively injective. Also, $e_i R e_{i+1}$ and $\widehat{K}$ are relatively injective.

Fact 2 $\text{Hom}(e_i R, \widehat{K}) = 0 = \text{Hom}(e_i R e_{i+1}, \widehat{K})$.

Fact 3 $e_i R$ and $e_j R$ are relatively injective for all $j \neq i$. Also, $e_i R e_{i+1}$ and $e_j R$ are relatively injective for all $j \neq i$.

Let U be an essential right ideal of R. Then $e_i R e_{i+1} \leq U$ for all $i = 1, 2, \ldots, n$. Set $W := \sum_{i=1}^{n} e_i R e_{i+1}$. Note that W is an ideal of R and $W \leq U$. Since the factor ring R/W is isomorphic to the ring $(\oplus_{i=1}^{n} D_i) \oplus \Delta$ and U/W is a right ideal of R/W, we conclude that there exists a partition I, J of the set $\{1, 2, \ldots, n\}$ and a right ideal K of Δ such that $U = (\oplus_{i \in I} e_i R) \oplus (\oplus_{j \in J} e_j R e_{j+1}) \oplus \widehat{K}$.

Now we deduce the following useful conclusions.

(i) $\oplus_{j \in J} e_j R e_{j+1}$ is a semisimple right R-module, and so $\oplus_{j \in J} e_j R e_{j+1}$ is quasi-injective.

(ii) $\oplus_{i \in I} e_i R$ is a quasi-injective right R-module. In fact, by Fact 3, we only need to prove that each $e_i R$ is a quasi-injective right R-module for all $i \in I$. Note that $e_i R e_{i+1}$ is only proper submodule of $e_i R$. Let $f : e_i R e_{i+1} \to$

$e_i R$ be an R-homomorphism. Note that $e_i R e_{i+1} = \begin{pmatrix} 0 & 0 & & & & \\ & 0 & 0 & & & \\ & & 0 & 0 & & \\ & & & \ddots & \ddots & V_i \\ & & & & \ddots & 0 & 0 \\ & & & & & & 0 \end{pmatrix}$.

Then $f(e_i R e_{i+1}) = e_i R e_{i+1}$. Since $dim(_{D_i}\{V_i\}) = dim(\{V_i\}_{D_{i+1}}) = 1$, there exists $v_i \in V_i$ such that $D_i v_i = v_i D_{i+1}$. Assume that $f(v_i) = \begin{pmatrix} 0 & 0 & & & & \\ & 0 & 0 & & & \\ & & 0 & 0 & & \\ & & \ddots & \ddots & v_i d_{i+1} & \\ & & & & \ddots & \\ & & & & 0 & 0 \\ & & & & & 0 \end{pmatrix}$ for some $d_{i+1} \in D_{i+1}$. There exists $d_i \in D_i$ such

that $d_i v_i = v_i d_{i+1}$. We consider the R-homomorphism $\bar{f} : e_i R \to e_i R$ defined

as left multiplication by $\begin{pmatrix} 0 & 0 & & & & \\ & 0 & 0 & & & \\ & & 0 & 0 & & \\ & & \ddots & d_i & 0 & \\ & & & & \ddots & \\ & & & & 0 & 0 \\ & & & & & 0 \end{pmatrix}$. Then $\bar{f}$ is an extension of f.

In the case of $e_n R$, it is similar.

(iii) $\widehat{K} = \widehat{K_1} \oplus \widehat{K_2}$, where $\widehat{K_1}$ is a quasi-injective R-module and $\widehat{K_2}$ is a square-free automorphism-invariant R-module. In fact, by Theorem 4.61, we have a decomposition $\Delta = \Delta_1 \times \Delta_2$, where Δ_1 is semisimple Artinian and Δ_2 is square-free. It follows that there exists a quasi-injective Δ-module K_1 and a square-free Δ-module K_2 such that $K = K_1 \oplus K_2$. Thus, $\widehat{K} = \widehat{K_1} \oplus \widehat{K_2}$. Since $e_{n+1} R(1 - e_{n+1}) = 0$, we obtain that $\widehat{K_1}$ is quasi-injective and $\widehat{K_2}$ is square-free by [10, lemma 2.3(6)]. Furthermore, by the hypothesis, $\widehat{K_2}$ is automorphism-invariant.

Let $X = (\oplus_{i \in I} e_i R) \oplus (\oplus_{j \in J} e_j R e_{j+1}) \oplus \widehat{K_1}$ and $Y = \widehat{K_2}$. Then $U = X \oplus Y$. By Facts 1, 2 and 3, X is quasi-injective, Y is automorphism-invariant square-free that is orthogonal to X, and X and Y are relatively injective. Then, clearly, U is automorphism-invariant. This shows that each essential right ideal of R is automorphism-invariant. Now, let A be any right ideal of R. Let C be a complement of A in R. Then $A \oplus C$ is an essential right ideal of R. Thus, as shown before, $A \oplus C$ is automorphism-invariant, and consequently, A is automorphism-invariant. This proves that R is a right a-ring. $\qquad \square$

Theorem 4.75 *Any indecomposable right Artinian right nonsingular right weakly CS right a-ring R is isomorphic to*

$$
\begin{pmatrix}
\mathbb{M}_{n_1}(e_1Re_1) & \mathbb{M}_{n_1 \times n_2}(e_1Re_2) & \mathbb{M}_{n_1 \times n_3}(e_1Re_3) & \cdots & \mathbb{M}_{n_1 \times n_k}(e_1Re_k) \\
0 & \mathbb{M}_{n_2}(e_2Re_2) & \mathbb{M}_{n_2 \times n_3}(e_1Re_2) & \cdots & \mathbb{M}_{n_2 \times n_k}(e_2Re_k) \\
0 & 0 & . & \cdots & . \\
\vdots & \vdots & \vdots & \vdots & \vdots \\
0 & 0 & 0 & \cdots & \mathbb{M}_{n_k}(e_kRe_k)
\end{pmatrix},
$$

where e_iRe_i is a division ring, $e_iRe_i \simeq e_jRe_j$ for each $1 \le i,j \le k$ and $n_1, \ldots, n_k$ are any positive integers. Furthermore, if $e_iRe_j \ne 0$, then

$$
dim_{(e_iRe_i}(e_iRe_j)) = 1 = dim((e_iRe_j)_{e_jRe_j}).
$$

Proof Let R be an indecomposable right Artinian right nonsingular right weakly CS right a-ring. We first show that eR is quasi-injective for any idempotent $e \in R$. Since R is right Artinian, we have $\text{Soc}(eR) \ne 0$. As R is right automorphism-invariant and right weak CS, eR is also a weak CS module. Therefore, $\text{Soc}(eR)$ is a simple module that is essential in eR, and so eR is uniform. Therefore, eR is quasi-injective.

Choose an independent family $\mathcal{F} = \{e_iR : 1 \le i \le n\}$ of indecomposable right ideals such that $R = \oplus_{i=1}^{n} e_iR$. After renumbering, we may write $R = [e_1R] \oplus [e_2R] \oplus \cdots \oplus [e_kR]$, where for $1 \le i \le k$, $[e_iR]$ denotes the direct sum of those e_jR that are isomorphic to e_iR. Let $[e_iR]$ be a direct sum of n_i copies of e_iR. Consider $1 \le i < j \le k$. We arrange the summands $[e_iR]$ in such a way that $l(e_jR) \le l(e_iR)$. Suppose $e_jRe_i \ne 0$. Then we have an embedding of e_iR into e_jR; hence, $l(e_iR) \le l(e_jR)$. But, by assumption, $l(e_jR) \le l(e_iR)$, so $l(e_iR) = l(e_jR)$, we get $e_jR \cong e_iR$, which is a contradiction. Hence, $e_jRe_i = 0$ for $j > i$. Thus, we have

$$
R \cong
\begin{bmatrix}
\mathbb{M}_{n_1}(e_1Re_1) & \mathbb{M}_{n_1 \times n_2}(e_1Re_2) & . & . & . & \mathbb{M}_{n_1 \times n_k}(e_1Re_k) \\
0 & \mathbb{M}_{n_2}(e_2Re_2) & . & . & . & \mathbb{M}_{n_2 \times n_k}(e_2Re_k) \\
0 & 0 & \mathbb{M}_{n_3}(e_3Re_3) & . & . & \mathbb{M}_{n_3 \times n_k}(e_3Re_k) \\
. & . & . & . & . & . \\
. & . & . & . & . & . \\
0 & 0 & . & & . & . & \mathbb{M}_{n_k}(e_kRe_k)
\end{bmatrix},
$$

where each $e_i Re_i$ is a division ring, $e_i Re_i \simeq e_j Re_j$ for each $1 \leq i, j \leq k$ and $n_1, \ldots, n_k$ are any positive integers. Furthermore, if $e_i Re_j \neq 0$, then

$$dim(_{e_i Re_i}(e_i Re_j)) = 1 = dim((e_i Re_j)_{e_j Re_j}). \qquad \qquad \square$$

Notes

The results of this chapter are taken from [79], [32], [94], [47] and [53].

5

Modules Coinvariant under Automorphisms of their Covers

We will devote this chapter to dualizing the results obtained in Chapter 2. Let M be a module and $\mathcal{X}$ be a class of R-modules closed under isomorphisms. A homomorphism $p \colon X \to M$ is an $\mathcal{X}$-*precover* if any other $g \colon X' \to M$ with $X' \in \mathcal{X}$ factorizes through it. And an $\mathcal{X}$-precover is called an $\mathcal{X}$-*cover* if, moreover, any $h \colon X \longrightarrow X$ such that $p \circ h = p$ must be an automorphism [137]. An $\mathcal{X}$-cover $p \colon X \longrightarrow M$ is called *epimorphic* if p is an epimorphism.

Definition 5.1 A module M with an $\mathcal{X}$-cover $p \colon X \to M$ is called $\mathcal{X}$-*endomorphism coinvariant* if, for any endomorphism $g \colon X \to X$, there exists an endomorphism $f \colon M \to M$ such that $f \circ p = p \circ g$.

Definition 5.2 A module M with an $\mathcal{X}$-cover $p \colon X \to M$ is called $\mathcal{X}$-*automorphism coinvariant* if, for any automorphism $g \colon X \to X$, there exists an endomorphism $f \colon M \to M$ such that $f \circ p = p \circ g$.

Remark 5.3 As in Remark 2.11, the preceding definition can be easily extended to modules having $\mathcal{X}$-precovers. Moreover, if $p \colon X \to M$ is an epimorphic cover, then M is $\mathcal{X}$-automorphism-invariant precisely when the cover p induces a group isomorphism $\Delta' \colon \mathrm{Aut}(M) \cong \mathrm{Aut}(X)/\mathrm{coGal}(X)$, where $\mathrm{coGal}(X) = \{g \in \mathrm{Aut}(X) \mid p \circ g = p\}$ is usually called the *co-Galois* group of the cover p (see e.g. [28, 29]).

Definition 5.4 A module M is called a quasi-projective module if, for every submodule N of M, any homomorphism $\varphi \colon M \to M/N$ can be lifted to a homomorphism $\psi \colon M \to M$.

Note that if $\mathcal{X}$ is the class of projective modules over a right perfect ring, then the $\mathcal{X}$-endomorphism-coinvariant modules are precisely the quasi-projective modules [136], and $\mathcal{X}$-automorphism coinvariant modules are called automorphism-coinvariant modules.

Example 5.5 Let R be the ring given in Example 4.21 and $M = e_{11}R$. As R is a finite-dimensional algebra over $\mathbb{F}_2$, the functors

$$\mathrm{Hom}_{\mathbb{F}_2}(-, \mathbb{F}_2) \colon \ \mathrm{Mod\text{-}R} \to \mathrm{R\text{-}Mod}$$

and

$$\mathrm{Hom}_{\mathbb{F}_2}(-, \mathbb{F}_2) \colon \ \mathrm{R\text{-}Mod} \to \mathrm{Mod\text{-}R}$$

establish a contravariant equivalence between the subcategories of left and right finitely generated modules over R. Moreover, as M is a finitely generated right R-module, its injective envelope $E(M)$ is also finitely generated.

Therefore, $\mathrm{Hom}_{\mathbb{F}_2}(E(M), \mathbb{F}_2)$ is the projective cover of $\mathrm{Hom}_{\mathbb{F}_2}(M, \mathbb{F}_2)$, and we deduce that $\mathrm{Hom}_{\mathbb{F}_2}(M, \mathbb{F}_2)$ is an automorphism-coinvariant left R-module. Moreover, it may be noticed that $\mathrm{Hom}_{\mathbb{F}_2}(M, \mathbb{F}_2)$ is not invariant under endomorphisms of its projective cover $\mathrm{Hom}_{\mathbb{F}_2}(E(M), \mathbb{F}_2)$ because otherwise $M \cong \mathrm{Hom}_{\mathbb{F}_2}(\mathrm{Hom}_{\mathbb{F}_2}(M, \mathbb{F}_2), \mathbb{F}_2)$ would be invariant under endomorphisms of $E(M) \cong \mathrm{Hom}_{\mathbb{F}_2}(\mathrm{Hom}_{\mathbb{F}_2}(M, \mathbb{F}_2), \mathbb{F}_2)$. Therefore, $\mathrm{Hom}_{\mathbb{F}_2}(M, \mathbb{F}_2)$ is not quasi-projective.

5.1 Structure and Properties

Notation 5.6 Throughout this section, $\mathcal{X}$ will be a class of modules closed under isomorphisms, and M with be a module with $p \colon X \to M$ an epimorphic $\mathcal{X}$-cover such that $\mathrm{End}(X)/J(\mathrm{End}(X))$ is a von Neumann regular, right self-injective ring and idempotents lift modulo $J(\mathrm{End}(X))$.

If $f \colon M \to M$ is an endomorphism, then there exists a $g \colon X \to X$ such that $p \circ g = f \circ p$. Moreover, if $g' \colon X \to X$ also satisfies that $p \circ g' = f \circ p$, then we get that $p \circ (g - g') = 0$. Thus, $p \circ (g - g') \circ t = 0$ for any $t \in S = \mathrm{End}(X)$, and this means that $p \circ (1 - (g - g') \circ t) = p$. Then, by the definition of cover, $1 - (g - g') \circ t$ is an automorphism. Thus, $1 - (g - g') \circ t$ is an isomorphism for all $t \in S$, and we get that $g - g' \in J(S)$. Therefore, we can define a ring homomorphism

$$\varphi \colon \ \mathrm{End}(M) \to S/J(S) \ \text{by} \ \varphi(f) = g + J(S).$$

Call $K = \mathrm{Ker}(\varphi)$ and $J = J(S)$. Then φ induces an injective ring homomorphism $\Psi \colon \mathrm{End}(M)/K \to S/J$. A dual argument to the one used in Lemma 3.2 proves the following.

Lemma 5.7 *Assume that $j \in J$. Then there exists an element $k \in K$ such that $p \circ j = k \circ p$.*

Theorem 5.8 *If M is $\mathcal{X}$-automorphism coinvariant, then* $\mathrm{End}(M)/K$ *is von Neumann regular,* $K = J(\mathrm{End}(M))$ *and idempotents lift modulo* $J(\mathrm{End}(M))$.

Proof Using Theorem 2.25, in order to show that $\mathrm{End}(M)/K$ is von Neumann regular, we only need to show that $\mathrm{Im}\,\Psi$ is invariant under left multiplication by units of S/J. This can be proved in a similar way as in Theorem 3.3.

As $\mathrm{End}(M)/K$ is von Neumann regular, clearly, $J(\mathrm{End}(M)) \subseteq K$. Let us prove the converse. As K is a two-sided ideal, we only need to show that $1 - f$ is invertible in $\mathrm{End}(M)$ for every $f \in K$. So take $f \in K$. Then $\Psi(1-f+K) = 1 + J$. Let $g\colon X \to X$ such that $(1 - f) \circ p = p \circ g$. Then $1 - g \in J$, and, thus, $g = 1 - (1 - g)$ is invertible. So there exists an $h\colon M \to M$ such, $p \circ g^{-1} = h \circ p$. Therefore,

$$h \circ (1 - f) \circ p = h \circ p \circ g = p \circ g^{-1} \circ g = p, (1 - f) \circ h \circ p$$
$$= (1 - f) \circ p \circ g^{-1} = p \circ g \circ g^{-1} = p.$$

And, as $p\colon X \to M$ is an epimorphic cover, we get that $(1 - f) \circ h$ and $h \circ (1-f)$ are automorphisms. Therefore, $1-f$ is invertible. So $K = J(\mathrm{End}(M))$.

Finally, the proof that idempotents lift modulo $J(\mathrm{End}(M))$ is also dual to the proof in Theorem 3.3 changing envelopes by covers.   □

The proofs similar to Corollary 3.8; Lemma 3.4 and Theorems 3.6, 3.9, 3.10 and 2.32 also show the following.

Corollary 5.9 *If M is $\mathcal{X}$-automorphism coinvariant, then M satisfies the finite exchange property.*

Lemma 5.10 *If M is $\mathcal{X}$-automorphism coinvariant and every direct summand of M has an $\mathcal{X}$-cover, then any direct summand of M is also $\mathcal{X}$-automorphism coinvariant.*

As discussed earlier in the paragraph before Theorem 3.6, for $M = N \oplus L$, we are again identifying $\mathrm{Hom}(N,L)$ and $\mathrm{Hom}(L,N)$ with appropriate subsets of $\mathrm{End}(M)$ in the next theorem.

Theorem 5.11 *If M is $\mathcal{X}$-automorphism coinvariant and every direct summand of M has an $\mathcal{X}$-cover, then M admits a decomposition $M = N \oplus L$ such that*

1. *N is a square-free module.*
2. *L is $\mathcal{X}$-endomorphism coinvariant, and $\mathrm{End}(L)/J(\mathrm{End}(L))$ is von Neumann regular, right self-injective and idempotents lift modulo $J(\mathrm{End}(L))$.*
3. *Both $\mathrm{Hom}_R(N,L)$ and $\mathrm{Hom}_R(L,N)$ are contained in $J(\mathrm{End}(M))$.*

In particular, $\mathrm{End}(M)/J(\mathrm{End}(M))$ *is the direct product of an abelian regular ring and a right self-injective von Neumann regular ring.*

Theorem 5.12 *Let* M *be an* $\mathcal{X}$*-automorphism-coinvariant module with epimorphic* $\mathcal{X}$*-cover* $p\colon X(M) \longrightarrow M$ *such that* $\mathrm{End}(X(M))/J(\mathrm{End}(X(M)))$ *is a von Neumann regular right self-injective ring and idempotents lift modulo* $J(\mathrm{End}(X(M)))$. *Assume that every direct summand of* M *has an* $\mathcal{X}$*-cover.*

1. *If, for each* $\mathcal{X}$*-endomorphism-coinvariant module, the finite exchange property implies the full exchange property, then* M *satisfies the full exchange property.*
2. M *is a clean module.*
3. *If* $\mathrm{End}(M)$ *has no homomorphic images isomorphic to* $\mathbb{Z}_2$, *then* M *is* $\mathcal{X}$*-endomorphism coinvariant and* $\mathrm{End}(M)/J(\mathrm{End}(M)) = \mathrm{End}(X)/J(\mathrm{End}(X))$. *In particular,* $\mathrm{End}(M)/J(\mathrm{End}(M))$ *is von Neumann regular, right self-injective and idempotents lift modulo* $J(\mathrm{End}(M))$.
   This is the case when $\mathrm{char}(\mathrm{End}(M)) = n > 2$ *and* $2 \nmid n$.

Lemma 5.13 *Let* M *be an* $\mathcal{X}$*-automorphism-coinvariant module with epimorphic* $\mathcal{X}$*-cover* $p\colon X(M) \longrightarrow M$ *such that* $\mathrm{End}(X(M))/J(\mathrm{End}(X(M)))$ *is a von Neumann regular right self-injective ring and idempotents lift modulo* $J(\mathrm{End}(X(M)))$. *If* M *is directly finite, then* $\mathrm{End}(M)/J(\mathrm{End}(M))$ *is unit regular.*

Theorem 5.14 *Let* M *be an* $\mathcal{X}$*-automorphism-coinvariant module with epimorphic* $\mathcal{X}$*-cover* $p\colon X(M) \longrightarrow M$ *such that* $\mathrm{End}(X(M))/J(\mathrm{End}(X(M)))$ *is a von Neumann regular right self-injective ring and idempotents lift modulo* $J(\mathrm{End}(X(M)))$. *Then the following are equivalent:*

1. M *is directly finite.*
2. M *has the internal cancellation property.*
3. M *has the cancellation property.*
4. M *has the substitution property.*
5. $X(M)$ *is directly finite.*
6. $X(M)$ *has the internal cancellation property.*
7. $X(M)$ *has the cancellation property.*
8. $X(M)$ *has the substitution property.*

Using similar arguments as in Lemma 2.35, Proposition 2.36 and Theorem 3.7, we can prove the following.

Lemma 5.15 *Let* M *be an* $\mathcal{X}$*-automorphism-coinvariant module with an epimorphic* $\mathcal{X}$*-cover* $p\colon X \to M$ *such that* $\mathrm{End}(X)/J(\mathrm{End}(X))$ *is von Neumann* *regular right self-injective and idempotents lift modulo $J(\text{End}(X))$. If X is indecomposable, then M is $\mathcal{X}$-endomorphism coinvariant.*

Proposition 5.16 *Let M be an indecomposable $\mathcal{X}$-automorphism-coinvariant module with an epimorphic $\mathcal{X}$-cover $p\colon X \to M$ such that $\text{End}(X)/J(\text{End}(X))$ is von Neumann regular right self-injective and idempotents lift modulo $J(\text{End}(X))$. Then the following statements are equivalent:*

1. *M is $\mathcal{X}$-endomorphism coinvariant.*
2. *X is indecomposable.*

Theorem 5.17 *Let M be an indecomposable $\mathcal{X}$-automorphism-coinvariant module with an epimorphic $\mathcal{X}$-cover $p\colon X \to M$ such that $\text{End}(X)/J(\text{End}(X))$ is von Neumann regular right self-injective and idempotents lift modulo $J(\text{End}(X))$. Assume that M is not $\mathcal{X}$-endomorphism coinvariant. Then $\text{End}(M)/J(\text{End}(M)) \cong \mathbb{F}_2$ and $\text{End}(X)/J(\text{End}(X))$ has a homomorphic image isomorphic to $\mathbb{F}_2 \times \mathbb{F}_2$.*

By arguments similar to Theorem 3.18 and Corollary 3.19, we may obtain the following dual results.

Theorem 5.18 *Let M be a module with an epimorphic $\mathcal{X}$-cover $p\colon X(M) \to M$ such that $\text{End}(X(M))/J(\text{End}(X(M)))$ is a von Neumann regular right self-injective ring and idempotents lift modulo $J(\text{End}(X(M)))$. Then the following are equivalent:*

1. *M is an $\mathcal{X}$-automorphism-coinvariant module.*
2. *M is an $\mathcal{X}$-epimorphism-coinvariant module.*

Corollary 5.19 *For a module M, we have the following:*

1. *M is automorphism coinvariant if and only if M is coinvariant under epimorphisms of its projective cover.*
2. *If M is cotorsion, then M is coinvariant under automorphisms of its flat cover if and only if it is coinvariant under all epimorphisms of its flat cover.*

Theorem 5.20 *Let $(\mathcal{F}, \mathcal{C})$ be a cotorsion pair such that $\mathcal{F}$ is closed under direct limits and every module has an $\mathcal{F}$-cover. If $M \in \mathcal{C}$ is $\mathcal{F}$-automorphism coinvariant, then*

1. *$\text{End}(M)/J(\text{End}(M))$ is von Neumann regular and idempotents lift modulo $J(\text{End}(M))$. Consequently, M satisfies the finite exchange property.*
2. *$M = N \oplus L$, where N is semi-Boolean and and L is $\mathcal{F}$-endomorphism coinvariant.*
3. *M is a clean module.*

5.2 Automorphism-Coinvariant Modules

When $\mathcal{X}$ is the class of projective modules, $\mathcal{X}$-automorphism-coinvariant modules are called automorphism-coinvariant modules. It is not difficult to see that automorphism-coinvariant modules generalize the class of projective modules and quasi-projective modules.

Theorem 5.21 *Let M be an automorphism-coinvariant module over a right perfect ring. Then*

1. $\text{End}(M)/J(\text{End}(M))$ *is von Neumann regular and idempotents lift modulo* $J(\text{End}(M))$.
2. $M = N \oplus L$, *where N is semi-Boolean and L is quasi-projective.*
3. *M satisfies the full exchange property.*
4. *M is clean.*
5. *If $\text{End}(M)$ has no homomorphic image isomorphic to $\mathbb{F}_2$, then M is quasi-projective.*

Proof (1) follows from Theorem 5.8, and (1) follows straight from Theorem 5.11. Theorem 5.21(3) follows from the fact that by (1), M satisfies the finite exchange property. Moreover, we know by (2) that $M = N \oplus L$, where N is semi-Boolean and L is quasi-projective. Thus, N satisfies the finite exchange property. Since N is square-free, this implies that N satisfies the full exchange property. It is known that a quasi-projective right module over a right perfect ring is discrete (see [83, theorem 4.41]). Thus, L is a discrete module. Since discrete modules satisfy the exchange property, we have that L satisfies the full exchange property. Thus, it follows that M satisfies the full exchange property. Theorem 5.21(4) follows from the fact that by Theorem 5.12(2), M is clean. Theorem 5.21(5) follows from Theorem 5.12(3). $\qquad\square$

Definition 5.22 A module M is called a pseudo-projective module if for every submodule N of M, any epimorphism $\varphi \colon M \to M/N$ can be lifted to a homomorphism $\psi \colon M \to M$.

Theorem 5.23 *If R is a right perfect ring, then a right R-module M is automorphism coinvariant if and only if it is pseudo-projective.*

Proof Let M be an automorphism-coinvariant module over a right perfect ring R with a projective cover $p \colon P \to M$. Let N be a submodule of M and $f \colon M \to M/N$, an epimorphism. It is known that $\text{End}(P)/J(\text{End}(P))$ is von Neumann regular, right self-injective and idempotents lift modulo $J(\text{End}(P))$. As projective modules are clearly automorphism coinvariant, by

Theorem 5.11, we have that $P = P_1 \oplus P_2$, where $\text{Hom}(P_1, P_2)$, $\text{Hom}(P_2, P_1) \subseteq J(\text{End}(P))$ and

$$\text{End}(P)/J(\text{End}(P)) = \text{End}(P_1)/J(\text{End}(P_1)) \times \text{End}(P_2)/J(\text{End}(P_2))$$

such that $\text{End}(P_1)/J(\text{End}(P_1))$ is a Boolean ring, and each element in the ring $\text{End}(P_2)/J(\text{End}(P_2))$ is the sum of two units.

Let us denote by $\pi: M \twoheadrightarrow M/N$ the canonical projection and call $S = \text{End}(P)$, $S_1 = \text{End}(P_1)$ and $S_2 = \text{End}(P_2)$. As $p: P \to M$ is a projective cover, there exists a direct summand $P(M/N)$ of P such that, if we denote by $v: P(M/N) \to P$ and $q: P \to P(M/N)$ the structural injection and projection, respectively, then $\pi \circ p \circ v: P(M/N) \to M/N$ is the projective cover of M/N, $e = v \circ q$ is an idempotent in S and $P(M/N) = eP$.

By projectivity, f lifts to an epimorphism $g: P \to P(M/N)$ such that $(\pi \circ p \circ v) \circ g = f \circ p$. This epimorphism g splits as $P(M/N)$ is projective. Thus, there exists a monomorphism $\delta: P(M/N) \to P$ such that $g \circ \delta = 1_{P(M/N)}$. Call $h = v \circ g$. As $h: P \to P$, $h \in S$. Note that $e \circ h = h$ and so $hS \subseteq eS$. Moreover, $h \circ \delta \circ q = v \circ g \circ \delta \circ q = v \circ q = e$. So $eS \subseteq hS$ and, consequently, $hS = eS$. This shows that $h: S \to eS$ is epic, and, consequently, $\bar{h}: \bar{S} \to \bar{e}\bar{S}$ is an epimorphism.

As $\bar{S} = \bar{S}_1 \times \bar{S}_2$, there exist idempotents $\bar{e}_1 \in \bar{S}_1$ and $\bar{e}_2 \in \bar{S}_2$, such that $\bar{e} = \bar{e}_1 \times \bar{e}_2 \in \bar{S}_1 \times \bar{S}_2$, and homomorphisms $\bar{h}_1: \bar{S}_1 \to \bar{S}_1$ and $\bar{h}_2: \bar{S}_2 \to \bar{S}_2$, such that $\bar{h} = \bar{h}_1 \times \bar{h}_2$.

Moreover, $\bar{e}_i \circ \bar{h}_i: \bar{S}_i \to \bar{e}_i\bar{S}_i$ is an epimorphism, and $(1 - \bar{e}_i) \circ \bar{h}_i = 0$ for $i = 1, 2$. As $\Im(\bar{h}) \cong \bar{e}\bar{S}$, it is a direct summand of $\bar{S}$. So there exists an idempotent $\bar{e}' \in \bar{S}$ such that $\Im(\bar{h}) = \bar{e}'\bar{S}$ and $\text{Ker}(\bar{h}) = (\bar{1} - \bar{e}')\bar{S}$. And, again, $\bar{e}' = \bar{e}'_1 \times \bar{e}'_2$ for idempotents $\bar{e}'_1 \in \bar{S}_1$ and $\bar{e}'_2 \in \bar{S}_2$. Also, we have $\text{Ker}(\bar{h}_1) = (1 - \bar{e}_1)S$ as $\bar{e}_1 \in \bar{S}_1$ is central because $\bar{S}_1$ is Boolean and $(1 - e)h = 0$. This yields $\bar{e}_1 = \bar{e}'_1$.

Call $\bar{h}'_1: \bar{S}_1 \to \bar{S}_1$ the homomorphism defined by $\bar{h}'_1|_{\bar{e}_1\bar{S}_1} = \bar{h}_1|_{\bar{e}_1\bar{S}_1}$ and $\bar{h}'_1|_{(1-\bar{e}_1)\bar{S}_1} = 1_{(1-\bar{e}_1)\bar{S}_1}$. By construction, $\bar{h}'_1$ is an automorphism in $\bar{S}_1$. On the other hand, $\bar{h}_2 \in \bar{S}_2$, so we can write $\bar{h}_2$ as the sum of two automorphisms, say $\bar{h}_2 = \bar{h}'_2 + \bar{h}''_2$. And, again, $\bar{h}''_2$ can be written as the sum of two automorphisms in $\bar{S}_2$, say $\bar{h}''_2 = \bar{t}_2 + \bar{t}'_2$.

Set $\bar{\gamma}_1 = \bar{h}'_1 \times \bar{h}'_2$, $\bar{\gamma}_2 = \bar{h}'_1 \times \bar{t}_2$ and $\bar{\gamma}_3 = (-\bar{h}'_1) \times \bar{t}'_2$. Consider then the homomorphism $\bar{\gamma} = \bar{\gamma}_1 + \bar{\gamma}_2 + \bar{\gamma}_3$. Then $\bar{\gamma}$ is the sum of three automorphisms $\bar{\gamma}_1$, $\bar{\gamma}_2$ and $\bar{\gamma}_3$ in $\bar{S}$. Note that for any $x_1 \times x_2 \in \bar{e}\bar{S} = \bar{e}_1\bar{S} \times \bar{e}_2\bar{S}$, we have that $(\bar{h}'_1 \times \bar{h}'_2 + \bar{h}'_1 \times \bar{t}_2 + (-\bar{h}'_1) \times \bar{t}'_2)(x_1 \times x_2) = \bar{h}'_1(x_1) \times \bar{h}_2(x_2) = \bar{h}_1(x_1) \times \bar{h}_2(x_2) = \bar{h}(x_1 \times x_2)$ since $\bar{h}'_1|_{\bar{e}_1\bar{S}} = \bar{h}_1|_{\bar{e}_1\bar{S}}$. This means that $\bar{\gamma}$ is the sum of three automorphisms in $\bar{S}$ and $\bar{\gamma}|_{\bar{e}\bar{S}} = \bar{h}|_{\bar{e}\bar{S}}$. Let us lift the three automorphisms $\bar{\gamma}_i \in \bar{S}$ to automorphisms $\gamma_i \in S$. As M is automorphism coinvariant, M is coinvariant under γ_i for $i = 1, 2, 3$. So if we call $\gamma = \gamma_1 + \gamma_2 + \gamma_3$, we get that M is coinvariant under γ. Moreover, as $\bar{e} \circ \bar{\gamma} = \bar{e} \circ \bar{h}$, there exists a $j \in J(S)$ such that $\pi \circ p \circ \gamma = \pi \circ p \circ h + \pi \circ p \circ j$. Thus, $\pi \circ p \circ h = \pi \circ p \circ (\gamma - j)$. As $j \in J(S)$, $1 - j$ is an automorphism and consequently, M is coinvariant under $1 - j$ and hence under j. We have already seen that M is coinvariant under γ. Thus, M is coinvariant under $\gamma - j$ and hence, by definition, it follows that there exists an endomorphism $t : M \to M$ such that $p \circ (\gamma - j) = t \circ p$. This gives $\pi \circ p \circ h = \pi \circ t \circ p$. As $\pi \circ p \circ h = f \circ p$, we have $\pi \circ t \circ p = f \circ p$. As p is epic, we have $\pi \circ t = f$.

Thus, t is an endomorphism of M such that $\pi \circ t = f$. This shows that M is pseudo-projective.

The converse is proved more generally later in Proposition 5.40. □

Remark 5.24 Note that the arguments given in the proof of Theorem 5.23 can be easily adapted to the situation in which the ring R is only assumed to be semiperfect instead of right perfect, and the module M is finitely generated, and, therefore, it has a projective cover.

We are now going to apply the preceding results to the special case in which $\mathcal{X}$ is the class of projective modules. Our first result is a dual of Theorem 4.23.

Theorem 5.25 *Let R be a right perfect ring and M be an automorphism-coinvariant right R-module with a projective cover $\pi : P \to M$ such that M is not quasi-projective. Assume that M is indecomposable with finite dual Goldie dimension. Then*

1. $\text{End}(M)/J(\text{End}(M)) \cong \mathbb{F}_2$.
2. *There exists an $n \geq 2$ and a set $\{P_i\}_{i=1}^n$ of non-isomorphic indecomposable projective modules such that $P = \oplus_{i=1}^n P_i$ and $\text{End}_R(P_i)/J(\text{End}(P_i)) \cong \mathbb{F}_2$ for every $i = 1, \ldots, n$.*

Proof By Lemma 5.17, $\text{End}(M)/J(\text{End}(M)) \cong \mathbb{F}_2$. By Theorem 5.11, we know that $M = N \oplus L$, where N is a semi-Boolean module and L is a quasi-projective module. As we are assuming that M is indecomposable and non quasi-projective, we deduce that $M = N$ is a semi-Boolean module. On the other hand, as R is right perfect, the projective cover P of M is a direct sum of indecomposable direct summands. Moreover, as M has finite dual Goldie dimension and M is square-free (because it is semi-Boolean and idempotents in $\text{End}_R(M)/J(\text{End}_R(M))$ lift to idempotents of $\text{End}(M)$), we deduce that there exists a finite set of non-isomorphic indecomposable projective modules $\{P_i\}_{i=1}^{n}$ such that $P = \oplus_{i=1}^{n} P_i$. Now $\operatorname{End}_R(P)/J(\operatorname{End}_R(P)) \cong \sqcap_{i=1}^{n} \operatorname{End}_R(P_i)/J(\operatorname{End}_R(P_i)) = \sqcap_{i=1}^{n} D_i$, where each $D_i = \operatorname{End}_R(P_i)/J(\operatorname{End}_R(P_i))$ is a division ring. By the proof of Theorem 5.11, we have that P is also a semi-Boolean module, and this means that any element in $\operatorname{End}_R(P)/J(\operatorname{End}_R(P))$ is idempotent and, therefore, each $D_i \cong \mathbb{F}_2$. Finally, $n \geq 2$ by Theorem 5.17. □

We can now adapt the arguments in Corollary 4.31 to first show the following.

Corollary 5.26 *Any automorphism-coinvariant module over a commutative local perfect ring is quasi-projective.*

This extends to any commutative perfect ring as shown in the following theorem.

Theorem 5.27 *Let R be a commutative perfect ring. Then any automorphism-coinvariant module over R is quasi-projective.*

Proof Let M be an automorphism-coinvariant module over R. Being a commutative (semi)perfect ring, R is a finite direct product of local rings, say $R = \sqcap_{i=1}^{n} R_i$. Then M is a direct product $M = \sqcap_{i=1}^{n} M_i$, where each M_i is an automorphism-coinvariant R_i-module. Therefore, each M_i is a quasi-projective R_i-module by the Corollary 5.26. And, thus, $M = \sqcap_{i=1}^{n} M_i$ is a quasi-projective $\sqcap_{i=1}^{n} R_i$-module. □

Remark 5.28 Observe that the same arguments used in the proofs show that the statements of Theorems 5.25 and 5.27 are also valid if we replace "(right) perfect ring" and "(right) module" with "semiperfect ring" and "finitely generated (right) module," respectively.

Let us finally note that Example 5.5 shows that over a noncommutative perfect ring, an automorphism-coinvariant module does not need to be quasi-projective.

5.3 Dual Automorphism-Invariant Modules

As we have seen in the previous section, the notion of automorphism coinvariance requires the existence of projective cover. Injective envelope of a module always exists, but the projective cover may or may not exist. Therefore, in order to deal with the situation when we do not necessarily have the existence of a projective cover, the notion of dual automorphism-invariant modules was introduced by Singh and Srivastava in [93].

Definition 5.29 A right R-module M is called a dual automorphism-invariant module if, whenever K_1 and K_2 are small submodules of M, then any epimorphism $\eta: M/K_1 \to M/K_2$ with small kernel lifts to an endomorphism φ of M.

We will show that, in fact, the endomorphism φ must be an automorphism of M. First, we have the following lemma.

Lemma 5.30 *Let M be a dual automorphism-invariant module. If $\sigma: M \to M$ is an epimorphism with small kernel, then σ is an automorphism.*

Proof Let $K = \mathrm{Ker}(\sigma)$. Then σ induces an isomorphism $\bar{\sigma}: M/K \to M$. Consider $\bar{\sigma}^{-1}: M \to M/K$. Since M is a dual automorphism-invariant module, by definition, $\bar{\sigma}^{-1}$ lifts to an endomorphism $\lambda: M \to M$. We have $\lambda(M) + K = M$. As $K \subset_s M$, we get $\lambda(M) = M$. Thus, λ is an epimorphism. Then, for any $x \in M$, $\bar{\sigma}^{-1}(x) = \lambda(x) + K$. Now $x = \bar{\sigma}\bar{\sigma}^{-1}(x) = \bar{\sigma}(\lambda(x) + K) = \sigma\lambda(x)$. This proves that $\sigma\lambda = 1_M$. Thus, $\sigma^{-1} = \lambda$ and hence σ is an automorphism. □

The following corollary follows as a consequence.

Corollary 5.31 *A right R-module M is a dual automorphism-invariant module if and only if, for any two small submodules K_1 and K_2 of M, any epimorphism $\eta: M/K_1 \to M/K_2$ with small kernel lifts to an automorphism φ of M.*

Proof Let M be a dual automorphism-invariant right R-module. Let K_1 and K_2 be any two small submodules of M, and let $\eta: M/K_1 \to M/K_2$ be any epimorphism with small kernel. Let $\ker(\eta) = L/K_1$. Then L is small in M. If $\pi: M \to M/K_1$ is a canonical epimorphism, then $\lambda = \eta\pi: M \to M/K_2$ has kernel L. Thus, $\lambda: M \to M/K_2$ is an epimorphism with small kernel. By definition, λ lifts to an endomorphism φ of M. Now $\varphi(M) + K_2 = M$. As $K_2 \subset_s M$, we get $\varphi(M) = M$. Thus, φ is an epimorphism with small kernel, and, hence, by Lemma 5.30, φ is an automorphism. The converse is obvious. □

Example 5.32 A module with no nonzero small submodule is easily seen to be a dual automorphism-invariant module. Thus, all the semiprimitive modules belong to the family of dual automorphism-invariant modules.

Definition 5.33 A ring R is called a right V-ring if every simple right R-module is injective.

The class of V-rings includes commutative von Neumann regular rings and von Neumann regular rings with Artinian primitive factors.

Proposition 5.34 *Let R be a right V-ring. Then every right R-module is dual automorphism invariant.*

Proof Let M be a nonzero right R-module. Let $x (\neq 0) \in M$. By the Zorn lemma, there exists a submodule N of M maximal with respect to not containing x. Then the intersection of all nonzero submodules of M/N is $(xR + N)/N$, and it is simple. Since R is a right V-ring, $(xR + N)/N$ is injective. Then $(xR + N)/N$ being a summand of M/N gives $M/N = (xR + N)/N$. Thus, $M = xR + N$. This shows that M has no nonzero small submodule, and, consequently, M is dual automorphism invariant. $\qquad\square$

In fact, the converse of the result also holds. To see this, we have the following useful observation.

Lemma 5.35 *Let M_1, M_2 be right R-modules. If $M = M_1 \oplus M_2$ is dual automorphism invariant, then any homomorphism $f : M_1 \to M_2/K_2$ with K_2 small in M_2 and $\mathrm{Ker}(f)$ small in M_1 lifts to a homomorphism $g : M_1 \to M_2$.*

Proof We have an epimorphism $\sigma : M \to M/K_2$ given by $\sigma(m_1 + m_2) = m_1 + f(m_1) + (m_2 + K_2)$ for $m_1 \in M_1, m_2 \in M_2$. Since K_2 is small in M_2 and $M_2 \subset M$, we get that K_2 is small in M. Now, as M is dual automorphism invariant, by definition, σ lifts to an automorphism η of M. Let $x_1 \in M_1$ and $\eta(x_1) = u_1 + u_2$, where $u_1 \in M_1, u_2 \in M_2$. Then $u_1 + u_2 + K_2 = (x_1 + K_2) + f(x_1)$, which gives $u_2 + K_2 = f(x_1)$. Let $\pi_2 : M \to M_2$ be the natural projection. Then $g = \pi_2 \eta|_{M_1} : M_1 \to M_2$ lifts f. $\qquad\square$

Now we have the following characterization of right V-rings in terms of dual automorphism-invariant modules.

Theorem 5.36 *A ring R is a right V-ring if and only if every finitely generated right R-module is dual automorphism invariant.*

Proof Suppose every finitely generated right R-module is dual automorphism invariant. We wish to show that R is a right V-ring. Assume to the contrary that R is not a right V-ring. Then there exists a simple right R-module S such that S is not injective. Let $E(S)$ be the injective hull of S. Then $E(S) \neq S$. Choose any $x \in E(S) \backslash S$. Then S is small in xR, and xR is uniform. Let $A = ann_r(x)$. As S is a submodule of $xR \cong R/A$, we may take $S = B/A$ for some $A \subset B \subset R_R$. Consider $M = R/A \times R/B$. As M is finitely generated, by hypothesis, M is dual automorphism invariant. We have the identity homomorphism $1_{R/B} : R/B \to R/B \cong (R/A)/(B/A)$, where $\mathrm{Ker}(1_{R/B}) = 0$ is small in R/B and B/A is small in R/A. By Lemma 5.35, the identity mapping on R/B can be lifted to a homomorphism $\eta: R/B \to R/A$. Thus, $\mathrm{Im}(\eta)$ is a summand of R/A, which is a contradiction to the fact that $R/A \ (\cong xR)$ is uniform. Hence, R is a right V-ring.

The converse is obvious from the Proposition 5.34. $\qquad\square$

Remark 5.37 It may be noted here that if we weaken the hypothesis and assume that R is a ring such that every cyclic right R-module is dual automorphism invariant, then R need not be a right V-ring. We know that every cyclic module over a commutative ring is quasi-projective, and it will be shown in Corollary 5.41 that every quasi-projective module with a projective cover is dual automorphism invariant. Thus, if we consider R to be a commutative perfect ring that is not Artinian, then every cyclic module over R is dual automorphism invariant, but R is not a V-ring.

Next, we will show that a module with a projective cover is dual automorphism invariant if and only if it is automorphism coinvariant. In order to show this, we have first the following useful lemma.

Lemma 5.38 *Let A, B be right R-modules, and let C be a small submodule of A. Let $f: A \to B$, and $g: A \to B$ be homomorphisms such that $g(C) = 0$. Consider induced homomorphisms $f': A \to B/f(C)$ and $g': A \to B/f(C)$. If $f' = g'$, then $f = g$.*

Proof Let $\pi: B \to B/f(C)$ be the natural projection. Then $f' = \pi f$ and $g' = \pi g$. Now, since $f' = g'$, we have for each $x \in A$, $f'(x) = g'(x)$. Thus, $\pi f(x) = \pi g(x)$ for each $x \in A$. This gives $f(x) + f(C) = g(x) + f(C)$ for each $x \in A$. So $(f - g)(x) \in f(C) = (f - g)(C)$ for each $x \in A$. Therefore, $(f - g)(A) \subseteq (f - g)(C)$, and, hence, $A \subseteq C + \mathrm{Ker}(f - g)$. Now, since C is a small submodule of A, we get $A = \mathrm{Ker}(f - g)$. Thus, $f - g = 0$, and, hence, $f = g$. $\qquad\square$

Now we are ready to prove the following proposition.

Proposition 5.39 *Let P be a projective module and $K \subset_s P$ such that $M = P/K$ is dual automorphism invariant. Then $\sigma(K) = K$ for any automorphism σ of P. Thus, any automorphism σ of P induces an automorphism of M. This implies, in particular, that every dual automorphism-invariant module with a projective cover is automorphism coinvariant.*

Proof Let $\sigma: P \to P$ be an automorphism. Suppose $\sigma(K) \not\subseteq K$. The map σ induces an epimorphism $\bar{\sigma}: P/K \to P/(K + \sigma(K))$. Its kernel is $(\sigma^{-1}(K) + K)/K$, which is small in $M = P/K$. Since M is dual automorphism invariant,

$\bar{\sigma}$ lifts to an automorphism η of M. Then η lifts to an automorphism λ of P. We have two epimorphisms, $\bar{\lambda}$ and μ, given as follows:

$$\bar{\lambda} : P \to P/K,$$

such that $\bar{\lambda}(x) = \lambda(x) + K = \eta(x + K)$, and

$$\mu : P \to P/K,$$

such that $\mu(x) = \sigma(x) + K$.

Let $\pi : P/K \to P/(K + \sigma(K))$ be a natural mapping. Then $\pi\bar{\lambda} = \pi\mu$, and, hence, by Lemma 5.38, we conclude that $\bar{\lambda} = \mu$. But $\text{Ker}(\mu) = \sigma^{-1}(K)$ and $\text{Ker}(\bar{\lambda}) = K$. This gives $\sigma^{-1}(K) = K$. Hence, $\sigma(K) = K$. $\qquad\square$

Theorem 5.40 *If a right R-module M has projective cover, then M is automorphism coinvariant if and only if M is dual automorphism invariant.*

Proof Suppose M has a projective cover and is automorphism coinvariant (equivalently, pseudo-projective). Let $M = P/K$, where P is projective and $K \subset_s P$. Let $\sigma : P \to P$ be an automorphism. Suppose $\sigma(K) \not\subset K$. We have an epimorphism

$$f : P \to P/K$$

given by $f(x) = \sigma(x) + K$.
We also have an epimorphism

$$h : P/K \to P/(K + \sigma(K))$$

given by $h(x + K) = \sigma(x) + K + \sigma(K)$.

As M is pseudo-projective, h lifts to an endomorphism β of M. Then, for $x + K \in P/K$, $h(x + K) = y + K + \sigma(K)$ whenever $\beta(x + K) = y + K$. As h is an epimorphism with kernel $(K + \sigma(K))/K$, we get $P/K = \beta(P/K) + (K + \sigma(K))/K$. Since $(K + \sigma(K))/K \subset_s P/K$, we have $P/K = \beta(P/K)$. Now $x \in \text{Ker}(\beta)$ implies $\sigma(x) + K + \sigma(K) = K + \sigma(K)$. Thus, $\sigma(x) \in K + \sigma(K)$, and, hence, $x \in \sigma^{-1}(K) + K$. As $\sigma^{-1}(K) + K$ is small, we get that $\text{Ker}(\beta)$ is small. Hence, β lifts to an automorphism γ of P and $\gamma(K) \subseteq K$.

Let $\pi : P \to P/K$ and $\pi_1 : P/K \to P/(K + \sigma(K))$ be canonical epimorphisms. Let $g = \pi\gamma$. Then $g(K) = \pi(\gamma(K)) = 0$ as $\gamma(K) \subseteq K = \text{Ker}(\pi)$. By the definition of γ, $h(x + K) = \gamma(x) + K + \sigma(K)$. Therefore, $\sigma(x) + K + \sigma(K) = \gamma(x) + K + \sigma(K)$, which gives us $\pi_1 g = \pi_1 f$. Hence, by Lemma 5.38, $f = g$. As $g(K) = 0$, we must have $f(K) = 0$. This gives $\sigma(K) \subseteq K$. Hence, M is dual automorphism invariant.

In Proposition 5.39, we have already seen that if M is a dual automorphism-invariant module with a projective cover, then M is automorphism coinvariant.
□

As a consequence, it follows that over a right perfect ring, the notions of automorphism coinvariance and dual automorphism invariance coincide.

Since every quasi-projective module is pseudo-projective, we have the following.

Corollary 5.41 *If a right R-module M has projective cover and is quasi-projective, then M is dual automorphism invariant.*

Next, we will show that dual automorphism-invariant modules need not be pseudo-projective. But, first we have the following useful observation.

Lemma 5.42 *Let M_1, M_2 be right R-modules. If $M = M_1 \oplus M_2$ is pseudo-projective, then M_1 is M_2 projective and M_2 is M_1 projective.*

Proof Let $f : M_1 \to M_2/N$ be a homomorphism. It induces an epimorphism $\sigma : M \to M/N$ given by $\sigma(x_1 + x_2) = x_1 + f(x_1) + (x_2 + N)$ for $x_1 \in M_1$, $x_2 \in M_2$. Since M is pseudo-projective, σ lifts to an endomorphism η of M. Let $x_1 \in M_1$ and $\eta(x_1) = u_1 + u_2$, where $u_1 \in M_1, u_2 \in M_2$. Then $u_1 + u_2 + N = x_1 + f(x_2) \in M_1 \oplus M_2/N$, $u_2 + N = f(x_2)$. Let $\pi_2 : M \to M_2$ be the natural projection. Then $\pi_2\eta|_{M_1} : M_1 \to M_2$ is such that $\pi_2\eta(x_1) = u_2$. This shows that $g = \pi_2\eta|_{M_1} : M_1 \to M_2$ lifts f. Hence, M_1 is M_2 projective. Similarly, it can be shown that M_2 is M_1 projective. □

Proposition 5.43 *If every right module over a ring R is pseudo-projective, then R is semisimple Artinian.*

Proof Let A be any right ideal of R. Since every right R-module is pseudo-projective, $R \oplus R/A$ is pseudo-projective. By Lemma 5.42, R/A is R-projective. Therefore, the identity mapping on R/A lifts to a mapping from R/A to R. Thus the exact sequence $0 \to A \to R \to R/A \to 0$ splits. Therefore, A is a summand of R. This shows that every right ideal of R is a summand of R. Hence, R is semisimple Artinian. □

Remark 5.44 *If R is a right V-ring that is not right Artinian (for example, a non-Artinian commutative von Neumann regular ring), then, by Proposition 5.34 and Proposition 5.43, it follows that R admits a dual automorphism-invariant module that is not pseudo-projective.*

Now we will discuss various properties of dual automorphism-invariant modules.

Proposition 5.45 *Any direct summand of a dual automorphism-invariant module is dual automorphism invariant.*

Proof Let M be a dual automorphism-invariant right R-module, and let $M = A \oplus B$. Let K_1, K_2 be two small submodules of A and $\sigma : A/K_1 \to A/K_2$ be an epimorphism. We get $\sigma' = \sigma \oplus 1_B : A/K_1 \oplus B \to A/K_2 \oplus B$. It lifts to an endomorphism η of M. For the inclusion map $\eta_1 : A \to M$ and the projection $\pi_1 : M \to A$, the map $\pi_1 \eta \eta_1 : A \to A$ lifts σ. Hence, A is dual automorphism invariant. This shows that any direct summand of a dual automorphism-invariant module is dual automorphism invariant. $\qquad \square$

As a consequence of the Lemma 5.35, we have the following:

Proposition 5.46 *If M_1, M_2 are two local modules such that $M_1 \oplus M_2$ is dual automorphism invariant, then M_1 is M_2 projective and M_2 is M_1 projective.*

Consider the following conditions on a module N:

($D1$): For every submodule A of N, there exists a decomposition
$N = N_1 \oplus N_2$ such that $N_1 \subseteq A$ and $N_2 \cap A \subset_s N$.
($D2$): If A is a submodule of N such that N/A is isomorphic to a direct
summand of N, then A is a direct summand of N.
($D3$): If A and B are direct summands of N with $A + B = N$, then
$A \cap B$ is a direct summand of N.

If N satisfies the condition ($D1$), then it is called a *lifting module*. If N satisfies the conditions ($D1$) and ($D3$), then it is called a *quasi-discrete module*. If N satisfies the conditions ($D1$) and ($D2$), then it is called a *discrete module*. The following implication is well known

$$\text{discrete} \implies \text{quasi-discrete} \implies \text{lifting}.$$

Since any quasi-projective module satisfies the property ($D2$) and hence the property ($D3$), it is natural to ask whether a dual automorphism-invariant module satisfies the property ($D2$). We do not know the answer to this question; however, we are able to show in the next proposition that every supplemented dual automorphism-invariant module satisfies the property ($D3$). Recall that a submodule K is called a *supplement* of N in M if $K + N = M$ and $K \cap N \subset_s M$. A module M is called a *supplemented module* if every submodule of M has a supplement.

Proposition 5.47 *If M is a supplemented dual automorphism-invariant module, then M satisfies the property ($D3$).*

Proof Let M be a supplemented dual automorphism-invariant module. Let A and B be direct summands of M such that $A + B = M$. We wish to show that $A \cap B$ is a direct summand of M. Since M is a supplemented module, there exists a submodule C of M such that $A \cap B + C = M$ and $A \cap B \cap C \subseteq_s M$. Now, clearly, we have $B = A \cap B + B \cap C$ and $A = A \cap B + A \cap C$. This gives $M = A \cap B + B \cap C + A \cap C$. Set $L = A \cap B \cap C$.

Now, as $C = A \cap C + B \cap C$, we have $L = A \cap B \cap (A \cap C + B \cap C) \subseteq_s M$. Thus,

$$\frac{M}{L} = \frac{A \cap B}{L} \oplus \frac{A \cap C}{L} \oplus \frac{B \cap C}{L}.$$

Since A is a direct summand of M, we have $M = A \oplus A'$ for some submodule A' of M. Then

$$\frac{M}{L} = \frac{A}{L} \oplus \frac{A' + L}{L} = \frac{A \cap B}{L} \oplus \frac{A \cap C}{L} \oplus \frac{A' + L}{L}.$$

Set $T = (A \cap B)/L \oplus (A' + L)/L$. Let $\pi : M/L \to T$ be the natural projection. Let us denote the restriction of π to T by π_T. Then $\pi_T : T \to T$ is an isomorphism. Thus, we have an isomorphism

$$1_{A \cap C/L} \oplus \pi_T : M/L \to M/L.$$

Since M is dual automorphism-invariant, this map lifts to an automorphism

$$\eta : M \to M.$$

We have

$$\eta(B) = (A \cap B) + (A' + L) = (A \cap B) + A' = (A \cap B) \oplus A'.$$

This shows that $A \cap B$ is a direct summand of $\eta(B)$. Now as $\eta(B)$ is a direct summand of M, we have that $A \cap B$ is a direct summand of M. Thus M satisfies the property $(D3)$.   □

We have already seen that if R is a right perfect ring, then dual automorphism-invariant modules over R are precisely the automorphism-coinvariant modules. Since every module over a right perfect ring is supplemented, it follows that every automorphism-coinvariant module over a right perfect ring satisfies the property $(D3)$.

Proposition 5.48 *Let R be a right perfect ring, and let M be a right R-module. If M is an automorphism-coinvariant lifting module, then M is discrete.*

Proof Let M be an automorphism-coinvariant lifting module. By Proposition 5.47, M satisfies the property $(D3)$. Thus, M is a quasi-discrete module with the property that every epimorphism $f \in \text{End}(M)$ with small kernel is an isomorphism. Hence, by [83, lemma 5.1], M is a discrete module.   □

Lemma 5.49 *Let R be a right perfect ring, and let $M = P/K$ be a right R-module where P is projective and $K \subset_s P$. Suppose M is a discrete module. Then*

1. *If P decomposes as $P = P_1 \oplus P_2$, then we get $M = M_1 \oplus M_2$ with*

$$M_1 = \frac{P_1 + K}{K}, M_2 = \frac{P_2 + K}{K};$$

and $K = K_1 \oplus K_2$ with

$$K_1 = K \cap P_1, K_2 = K \cap P_2.$$

This shows any decomposition of P gives rise to natural decompositions of both M and K.

2. *If $\sigma \in \text{End}(P)$ is an idempotent, then $\sigma(K) \subseteq K$.*

Proof 1. Let $P = P_1 \oplus P_2$. Then $M = (P_1 + K)/K + (P_2 + K)/K$. Let L_1 and L_2 be projections of K in P_1 and P_2 respectively. Then L_1, L_2, $L_1 + L_2$ are small in M. Now $P_1 \cap (P_2 + K) \subseteq L_1 + L_2$, so $(P_1 + K) \cap (P_2 + K) \subseteq K + L_1 + L_2$. This gives that $((P_1 + K)/K) \cap ((P_2 + K)/K) \subset_s M$. Since M satisfies the property $(D1)$, we get that $M = A/K + B/K$ such that $A/K, B/K$ are summands of M contained in $(P_1 + K)/K$, $(P_2 + K)/K$, respectively, and are supplements of $(P_2 + K)/K$, $(P_1 + K)/K$, respectively. As M satisfies the property $(D3)$, $A/K \cap B/K$ is a summand of M. However, $A/K \cap B/K \subseteq ((P_1 + K)/K) \cap ((P_2 + K)/K)$ gives that $A/K \cap B/K$ is small in M. Therefore, $M = A/K \oplus B/K$ and hence

$$M = \frac{P_1 + K}{K} \oplus \frac{P_2 + K}{K}.$$

Let $K_1 = K \cap P_1, K_2 = K \cap P_2$. We have an isomorphism $\varphi: P_1/K_1 \oplus P_2/K_2 \to M$ given by $\varphi(x_1 + K_1, x_2 + K_2) = x_1 + x_2 + K$, where $x_1 \in P_1, x_2 \in P_2$. As $\varphi(K/(K_1 + K_2)) = 0$, we get $K = K_1 \oplus K_2$. Hence, $K = (K \cap P_1) \oplus (K \cap P_2)$.

2. Let $P_1 = \sigma P$, and $P_2 = (1 - \sigma)P$. Then $P = P_1 \oplus P_2$. By (1), we have $K = K_1 \oplus K_2$ where $K_1 = K \cap P_1, K_2 = K \cap P_2$. Clearly, then, $\sigma(K) \subseteq K$.   □

Next, we consider automorphism-coinvariant lifting modules over right perfect rings.

Theorem 5.50 *Let R be a right perfect ring, and let M be a right R-module. If M is an automorphism-coinvariant lifting module, then M is quasi- projective.*

Proof Let $M = P/K$, where P is projective and $K \subset_s P$. Let $\sigma \in End(P)$. We know that $End(P)$ is clean. Therefore, $\sigma = \alpha + \beta$, where α is an idempotent in $End(P)$ and β is an automorphism on P. Since M is an automorphism-coinvariant lifting module over a right perfect ring, by Proposition 5.48, M is discrete. Therefore, by Lemma 5.49(2), $\alpha(K) \subseteq K$. Since M is an automorphism-coinvariant module, by Theorem 5.39, $\beta(K) \subseteq K$. Thus, $\sigma(K) = (\alpha + \beta)(K) \subseteq K$. Hence, M is quasi- projective. $\qquad\square$

Theorem 5.51 *Let R be a right perfect ring. If $M = M_1 \oplus M_2$ is a dual automorphism-invariant right R-module, then both M_1 and M_2 are dual automorphism-invariant and they are projective relative to each other.*

Proof We have already seen that a direct summand of a dual automorphism-invariant module is dual automorphism-invariant. Now, we proceed to show that M_1 and M_2 are projective relative to each other. Let $M_1 = P_1/K_1$ and $M_2 = P_2/K_2$ where P_1, P_2 are projective and $K_1 \subset_s P_1, K_2 \subset_s P_2$. Then $M = M_1 \oplus M_2 = (P_1 \oplus P_2)/(K_1 \oplus K_2)$. Note that the decomposition $M = M_1 \oplus M_2$ gives rise to decomposition $P = P_1 \oplus P_2$, where $M_1 = (P_1 + K)/K$ and $M_2 = (P_2 + K)/K$ and $K = K_1 \oplus K_2$, where $K_1 = K \cap P_1, K_2 = K \cap P_2$. Thus, $M_1 \cong P_1/K_1$ and $M_2 \cong P_2/K_2$.

Let $\overline{L_2} = L_2/K_2$ be any submodule of M_2. Consider the exact sequence $M_2 \to M_2/\overline{L_2} \to 0$. Let $\lambda: M_1 \to M_2/\overline{L_2}$ be a homomorphism. This gives us a mapping

$$\lambda': P_1/K_1 \to P_2/L_2$$

with $\lambda'(x_1 + K_1) = x_2 + L_2$ if $\lambda(x_1 + K_1) = (x_2 + K_2) + L_2/K_2$.

It lifts to a homomorphism $\mu: P_1 \to P_2$. Then $P = P_1' \oplus P_2$ where $P_1' = \{x_1 + \lambda'(x_1): x_1 \in P_1\}$. We get an automorphism

$$\sigma: P \to P,$$

where $\sigma(x_1 + x_2) = x_1 + \lambda'(x_1) + x_2$.

Since M is dual automorphism-invariant, we have $\sigma(K) = K = \sigma(K_1) \oplus \sigma(K_2) = K \cap P_1' \oplus K \cap P_2$. This gives a decomposition

$$M = \frac{P_1' + K}{K} \oplus \frac{P_2 + K}{K}.$$

We have an isomorphism

$$\sigma' : \frac{P_1 + K}{K} \to \frac{P_1' + K}{K}$$

given by $\sigma'(x_1 + K) = \sigma(x_1) + K = x_1 + \lambda'(x_1) + K$. Now, if $x_1 \in K$, then $x_1 + \lambda'(x_1) \in K$. This gives $\lambda'(x_1) \in K \cap P_2 = K_2$. Hence, λ' induces mapping

$$\bar{\mu} : \frac{P_1}{K_1} \to \frac{P_2}{K_2}$$

given by $\bar{\mu}(x + K_1) = \lambda'(x) + K_2$. This shows that M_1 is projective with respect to M_2. Similarly, it can be shown that M_2 is projective with respect to M_1. $\square$

As a consequence follows Corollary 5.52.

Corollary 5.52 *If R is a right perfect ring, then a right R-module M is quasi-projective if and only if $M \oplus M$ is dual automorphism-invariant.*

Proof Let R be a right perfect ring. Suppose M is a quasi-projective right R-module. Then $M \oplus M$ is quasi-projective and, hence, dual automorphism-invariant. Conversely, suppose $M \oplus M$ is dual automorphism-invariant. Then, by Theorem 5.51, M is M-projective; that is, M is quasi-projective. $\square$

Proposition 5.53 *Let R be an Artinian serial ring. If M is dual automorphism-invariant, then M is quasi-projective.*

Proof Since R is Artinian serial, $M = \oplus_{i=1}^{n} M_i$, where each M_i is uniserial. Since M is dual automorphism-invariant, by Theorem 5.51, each M_i is projective with respect to M_j, for each $j \neq i$.

Let M_i, M_j be such that $M_i/(M_i J(R)) \cong M_j/(M_j J(R))$. Then, since M_i, M_j are projective relative to each other, we can lift this isomorphism to give $M_i \cong M_j$.

So now $M = \oplus_{i=1}^{m} L_i$, where $L_i = \oplus_{i \in \Lambda} M_i$ with $M_i \cong M_k$ for each $i, k \in \Lambda$. Let t be the length of $M_i \subset L_i$. Then, as R is an Artinian serial ring, M_i is projective as an $R/J^t(R)$-module for each $i \in \Lambda$. This shows that L_i is M-projective. Consequently, it follows that M is M-projective. Thus, M is quasi-projective. $\square$

Next, we discuss dual automorphism-invariant abelian groups.

Proposition 5.54 *Let P be a projective right R-module that has no nonzero small submodule, and let M be any quasi-projective right R-module such that*

$\text{Hom}_R(M/K, P) = 0$ *for any small submodule K of M. Then* $P \oplus M$ *is dual automorphism-invariant.*

Proof Set $N = P \oplus M$. We have projections $\pi_1 : N \to P$, and $\pi_2 : N \to M$. Let K be a small submodule of N. Then $\pi_1(K) \subset_s P$. This gives $\pi_1(K) = 0$ as P has no nonzero small submodule. Therefore, $K \subset M$. Let K_1, K_2 be two small submodules of N. Then $N/K_1 = P \oplus (M/K_1)$ and $N/K_2 = P \oplus (M/K_2)$.

Let $\sigma : N/K_1 \to N/K_2$ be an epimorphism. Now σ may be viewed as $\sigma = \begin{bmatrix} \sigma_{11} & \sigma_{12} \\ \sigma_{21} & \sigma_{22} \end{bmatrix}$, where $\sigma_{11} : P \to P$, $\sigma_{12} : M/K_1 \to P$, $\sigma_{21} : P \to M/K_2$, $\sigma_{22} : M/K_1 \to M/K_2$.

Set $\lambda_{11} = \sigma_{11}$, $\lambda_{12} : M \to P$ naturally given by σ_{12}, and $\lambda_{21} : P \to M$ a lifting of σ_{21}. As M is quasi-projective, σ_{22} lifts to an endomorphism λ_{22} of M. Let $\eta = \begin{bmatrix} \lambda_{11} & \lambda_{12} \\ \lambda_{21} & \lambda_{22} \end{bmatrix}$. Then η is an endomorphism of N. As $\lambda_{12} = 0$ by the hypothesis, for any $x \in K_1$, $\begin{bmatrix} \lambda_{11} & \lambda_{12} \\ \lambda_{21} & \lambda_{22} \end{bmatrix} \begin{bmatrix} 0 \\ x \end{bmatrix} = \begin{bmatrix} \lambda_{12}(x) \\ \lambda_{22}(x) \end{bmatrix} = \begin{bmatrix} 0 \\ \lambda_{22}(x) \end{bmatrix}$. As λ_{22} is a lifting of σ_{22}, $\lambda_{22}(K_1) \subseteq K_2$. Hence, η lifts σ. This proves that $P \oplus M$ is dual automorphism-invariant. $\quad\square$

In particular, for abelian groups, we have the following corollary.

Corollary 5.55 *Let P be a projective abelian group and let M be any torsion quasi-projective abelian group. Then* $P \oplus M$ *is dual automorphism-invariant.*

Proof As P is a direct sum of copies of $\mathbb{Z}$ and $\mathbb{Z}$ has no nonzero small subgroup, P has no nonzero small subgroup. It is easy to observe that $\text{Hom}_{\mathbb{Z}}(M, P) = 0$. Thus, the result follows from the preceding lemma. $\quad\square$

The preceding result gives us plenty of examples of dual automorphism-invariant modules.

Example 5.56 Let $M = \mathbb{Z} \oplus C$, where C is a finite cyclic group. By Corollary 5.55, M is dual automorphism-invariant but M is not pseudo-projective unless $C = 0$.

An abelian group G is said to be of *bounded index* if there exists a bound on the orders of elements of G. A nonzero element $x \in G$ is said to have *finite height* if there exists a bound on the orders of cyclic subgroups of G that contain x. An abelian group G is called a *divisible group* if for each positive integer n and every element $g \in G$, there exists $h \in G$ such that $nh = g$. An abelian group G is called a *reduced group* if G has no proper divisible subgroup. We recall some useful facts from the theory of abelian groups and summarize later. See Fuchs [36] for further details on the theory of abelian groups.

Theorem 5.57 *(See [36]). Let G be an abelian p-group.*

1. *If G is of bounded index, then G is a direct sum of cyclic p-groups.*
2. *If $x (\neq 0) \in Soc(G)$ has finite height and if H is a cyclic subgroup of G of largest order that contains x, then H is a summand of G.*
3. *If every element in the socle of G is of infinite height, then G is divisible.*

Thus, it follows that if a torsion abelian group is not divisible, then it admits a cyclic summand.

Theorem 5.58 *(See [36].) If an abelian group G is such that its torsion subgroup H has a bound on orders of its elements, then the torsion subgroup H is a summand of G.*

Theorem 5.59 *(See [36].) If G is an abelian group, then $G = D \oplus K$, where D is divisible and K is reduced. Furthermore, $D \cong \oplus_{m_p} \mathbb{Z}(p^\infty) \oplus_n \mathbb{Q}$.*

We have the following observation for a torsion abelian group.

Lemma 5.60 *Let G be a torsion abelian group such that G is dual automorphism-invariant. Then, G cannot be divisible.*

Proof Assume to the contrary that G is divisible. Then, in view of Theorem 5.59, we have $G \cong \oplus_{m_p} \mathbb{Z}(p^\infty)$. For a prime number p, consider $H = \mathbb{Z}(p^\infty)$. In it, every proper subgroup is small. Let $A \subsetneq B$ be two proper subgroups of H. There exists an isomorphism $\sigma : H/A \to H/B$. Since every summand of a dual automorphism-invariant module is dual automorphism-invariant, H is dual automorphism-invariant. Therefore, σ lifts to an endomorphism η of H. Then $\sigma(A) = B$. This gives a contradiction as order of A is less than the order of B. Hence, G cannot be divisible. $\square$

Next, we recall the characterization of quasi-projective abelian groups due to Fuchs and Rangaswamy.

Theorem 5.61 *An abelian group G is quasi-projective if and only if it is either free or a torsion group such that every p-component G_p is a direct sum of cyclic groups of the same order p^n.*

Now we are ready to prove the following for a torsion dual automorphism-invariant abelian group.

Proposition 5.62 *Let G be a torsion dual automorphism-invariant abelian group. Then G is quasi-projective, and it has no divisible subgroup.*

Proof Since any abelian group is a direct sum of a divisible group and a reduced group, in view of Lemma 5.60, it follows that G is reduced. Let p

be a prime number. Consider the p-component G_p of G. Suppose $G_p \neq 0$. As G_p is reduced, $G_p = A_1 \oplus L$, where A_1 is a nonzero cyclic p-group. Now $o(A_1) = p^n$ for some $n > 0$. If $L = 0$, we get that G_p is projective. Suppose $L \neq 0$. Then $L = A_2 \oplus L_1$, where A_2 is a nonzero cyclic p-group. By Proposition 5.45, $A_1 \oplus A_2$ is dual automorphism-invariant. As every subgroup of A_1 or A_2 is small, it follows that A_1 is A_2-projective and A_2 is A_1-projective. Hence, $A_1 \oplus A_2$ is quasi-projective. This gives $A_1 \cong A_2$. By the preceding theorem, we get G_p is a direct sum of copies of A_1. Hence, G_p is quasi-projective. This proves that G itself is quasi-projective. $\square$

Lemma 5.63 *Let G be a torsion-free, uniform abelian group that is not finitely generated. Let H be a nontrivial cyclic subgroup of G. For any prime number p, let $G_p = \{x \in G : p^n x \in H \text{ for some } n \geq 0\}$. Then $J(G) \neq 0$ if and only if the number of prime numbers p for which $G_p = H$ is finite.*

Proof Observe that $H \subseteq G_p$ for any prime number p. Without loss of generality, we take $G \subseteq \mathbb{Q}$ and $H = \mathbb{Z}$. Let M be a maximal subgroup of G. For some prime number p, G/M is of order p. Thus, $pG_p \subseteq M$. Now G_p is generated by some powers $1/p^n$, $n \geq 0$.

Case 1. Assume $\mathbb{Z} \subset G_p$. Then $\mathbb{Z} \subseteq pG_p \subseteq M$, $M/\mathbb{Z}$ is a maximal subgroup of $G/\mathbb{Z}$. As $G/\mathbb{Z}$ is a torsion group such that for each prime number q, $G_q/\mathbb{Z}$ is the q-torsion component of $G/\mathbb{Z}$, we get $G_q \subseteq M$, whenever $q \neq p$. Then $M = (G_p \cap M) + A_p$, where A_p is the sum of all G_q, $q \neq p$.

Case 2. Assume $\mathbb{Z} = G_p$. If $\mathbb{Z} \subseteq M$, the arguments of Case 1 show that $M = G$, which is a contradiction. Thus, $\mathbb{Z} \not\subseteq M$, and we get $M \cap \mathbb{Z} = p\mathbb{Z}$.

We know that the intersection of infinitely many sets $p\mathbb{Z}$ is zero. Thus, it follows that $J(G) \neq 0$ if and only if the number of primes p for which $G_p = \mathbb{Z}$ is finite. $\square$

Theorem 5.64 *Let G be a subgroup of $\mathbb{Q}$ containing $\mathbb{Z}$. Then the following conditions are equivalent:*

1. *G is dual automorphism-invariant.*
2. *The number of primes p for which $G_p = \{x \in G : p^n x \in \mathbb{Z}\} = \mathbb{Z}$ is not finite.*
3. *$J(G) = 0$.*

Proof (1) $\implies$ (2). Let G be a subgroup of $\mathbb{Q}$ containing $\mathbb{Z}$, and suppose G is dual automorphism-invariant. Assume to the contrary that the number of primes p for which $G_p = \{x \in G : p^n x \in \mathbb{Z}\} = \mathbb{Z}$ is not finite. Then by Lemma 5.63, $J(G) \neq 0$. Therefore, we can find a cyclic subgroup H that is small. We take $H = \mathbb{Z}$. By using the Lemma 5.63, we see that $G/\mathbb{Z}$ is an infinite direct sum of its p-components. For any prime number $p \neq 2$ for which the p-component $G_p/\mathbb{Z}$ is nonzero, its group of automorphisms is of order more than one. This proves that $Aut(G/\mathbb{Z})$ is uncountable. As $\mathbb{Q}$ is countable, it follows that some automorphism of $G/\mathbb{Z}$ cannot be lifted to endomorphism of G. Hence, G is not dual automorphism-invariant, which is a contradiction. This proves that the number of primes p for which $G_p = \{x \in G: p^n x \in \mathbb{Z}\} = \mathbb{Z}$ is not finite.

$(2) \Longrightarrow (3)$. It follows from Lemma 5.63.

$(3) \Longrightarrow (1)$ is trivial.　　　　　　　　　　　　　　　　　□

Corollary 5.65 *If a torsion-free abelian group G is divisible, then it is not dual automorphism-invariant.*

Proof Let G be a torsion-free divisible abelian group. Assume to the contrary that G is dual automorphism-invariant. As $\mathbb{Q}$ is a summand of G, it must be dual automorphism-invariant by Proposition 5.45. However, we know that $\mathbb{Q}$ is not dual automorphism-invariant (see Theorem 5.64). This yields a contradiction. Hence, G is not dual automorphism-invariant.　　　　□

From Theorem 5.59, Lemma 5.60 and Corollary 5.65, we conclude the following theorem.

Theorem 5.66 *Let G be a dual automorphism-invariant abelian group. Then G cannot be divisible, and hence, G must be reduced.*

Notes

The results of this chapter are taken from [49], [93] and [53].

6

Schröder–Bernstein Problem

The Schröder–Bernstein theorem is a classical result in basic set theory. It states that if A and B are two sets such that there is a one-to-one function from A into B and a one-to-one function from B into A, then there exists a bijective map between the two sets A and B. This type of problem where one asks if two mathematical objects A and B that are similar in some sense to part of each other are also similar is usually called the Schröder–Bernstein problem, and it has been studied in various branches of mathematics.

In the context of modules, this problem was studied by Bumby in [14], where he proved that the Schröder–Bernstein problem has a positive solution for modules that are invariant under endomorphisms of their injective envelope. To prove it, he first showed that if M and N are two modules such that there is a monomorphism from M to N and a monomorphism from N to M, then their injective envelopes are isomorphic; that is, $E(M) \cong E(N)$. As a consequence, he deduced that if M and N are two modules invariant under endomorphisms of their injective envelopes such that there is a monomorphism from M to N and a monomorphism from N to M, then $M \cong N$.

6.1 Schröder–Bernstein Problem for $\mathcal{X}$-Endomorphism Invariant Modules

Let $\mathcal{X}$ be an enveloping class of right R-modules. We will assume that every right R-module M has a monomorphic $\mathcal{X}$-envelope that we are going to denote by $v_M \colon M \to X(M)$.

Definition 6.1 A homomorphism $u \colon N \to M$ of right R-modules is called an $\mathcal{X}$-strongly pure monomorphism if any homomorphism $f \colon N \to X$, with $X \in \mathcal{X}$, extends to a homomorphism $g \colon M \to X$ such that $g \circ u = f$.

141

Let us note that $\mathcal{X}$-strongly pure monomorphisms are clearly closed under composition. Moreover, if $X \in \mathcal{X}$, then any $\mathcal{X}$-strongly pure monomorphism $u : X \to M$ splits.

The following characterization of $\mathcal{X}$-strongly pure monomorphisms is straightforward.

Lemma 6.2 *Let* $u : N \to M$ *be a homomorphism. Then the following are equivalent.*

1. *u is an $\mathcal{X}$-strongly pure monomorphism.*
2. *$v_N : N \to X(N)$ factors through u.*
3. *The composition $v_M \circ u : N \to X(M)$ is an $\mathcal{X}$-preenvelope.*

Observe that condition (2) implies that any $\mathcal{X}$-strongly pure monomorphism is a monomorphism, as we are assuming that every module has a monomorphic $\mathcal{X}$-envelope.

Definition 6.3 A submodule N of M is called an $\mathcal{X}$-strongly pure submodule if the inclusion map $i : N \to M$ is an $\mathcal{X}$-strongly pure monomorphism.

Given a right R-module M, we will denote by add[M] the class of all direct summands of finite direct sums of copies of M.

Definition 6.4 A module M is called $\mathcal{X}$-strongly purely closed if any direct limit of splitting monomorphisms among objects in add[M] is an $\mathcal{X}$-strongly pure monomorphism.

We give some examples of $\mathcal{X}$-strongly purely closed modules.

Example 6.5 1. Let $\mathcal{X}$ be the class of all injective modules. Then any module is $\mathcal{X}$-strongly purely closed.
2. Let $\mathcal{X}$ be the class of all pure-injective modules. Then any module is $\mathcal{X}$-strongly purely closed.
3. Let $(\mathcal{F},\mathcal{C})$ be a cotorsion pair cogenerated by a set (see [37]), and assume that $\mathcal{F}$ is closed under taking direct limits. Then it is known that every module has a monomorphic $\mathcal{C}$-envelope (see e.g. [137]). It is easy to check that any object in $\mathcal{F} \cap \mathcal{C}$ is $\mathcal{C}$-strongly purely closed.

In the following proposition, we describe the endomorphism ring of $\mathcal{X}$-strongly purely closed modules.

Proposition 6.6 *Let $\mathcal{X}$ be a class of modules closed under isomorphisms, and assume any module has a monomorphic $\mathcal{X}$-envelope. Then, for any $\mathcal{X}$-strongly purely closed module X, $\mathrm{End}(X)$ is a right cotorsion ring.*

In particular, $\mathrm{End}(X)/J(\mathrm{End}(X))$ *is von Neumann regular right self-injective and idempotents lift modulo* $J(\mathrm{End}(X))$.

Proof Let us call $S = \mathrm{End}(X)$. Take any short exact sequence $0 \to S_S \xrightarrow{u} L \to F \to 0$ with F, a flat right S-module. As F is flat, the sequence is pure, and thus, the induced sequence $0 \to S_S \otimes X_R \to L \otimes X_R \to F \otimes X_R \to 0$ is also pure in Mod-R. On the other hand, we know that F is a direct limit of a family of finitely generated projective modules. Say that $F = \varinjlim P_i$. Let us denote by $\delta_i : P_i \to F$ the canonical homomorphisms from P_i to the direct limit. Taking pullbacks, we get the commutative diagram,

$$\begin{array}{ccccccccc} 0 & \longrightarrow & S & \xrightarrow{u_i} & L_i & \longrightarrow & P_i & \longrightarrow & 0 \\ & & \downarrow{\cong} & & \downarrow{\varphi_i} & & \downarrow{\delta_i} & & \\ 0 & \longrightarrow & S & \xrightarrow{u} & L & \longrightarrow & F & \longrightarrow & 0, \end{array}$$

in which the upper row splits, since P_i is projective. Moreover, $L = \varinjlim L_i$. Now applying the functor $- \otimes_S X$, we get the following commutative diagram in Mod-R:

$$\begin{array}{ccccccccc} 0 & \longrightarrow & S \otimes_S X & \xrightarrow{u_i \otimes X} & L_i \otimes X & \longrightarrow & P_i \otimes X & \longrightarrow & 0 \\ & & \downarrow{1_S \otimes X} & & \downarrow{\varphi_i \otimes X} & & \downarrow{\delta_i \otimes X} & & \\ 0 & \longrightarrow & S \otimes X & \xrightarrow{u \otimes X} & L \otimes X & \longrightarrow & F \otimes X & \longrightarrow & 0. \end{array}$$

We have $L \otimes X = \varinjlim L_i \otimes X$ and $F \otimes X = \varinjlim P_i \otimes X$, since $- \otimes_S X$ commutes with direct limits. Note that $S \otimes_S X \cong X$ and $P_i \otimes X$ is isomorphic to a direct summand of a finite direct sum of copies of X. This means that $u \otimes X$ is a direct limit of splitting monomorphisms among modules in add[X], and as we are assuming that X is an X-strongly purely closed module, this means that $u \otimes X$ is an X-strongly pure monomorphism. So there exists an $h : L \otimes_S X \to S \otimes X$ such that $h \circ (u \otimes X) = 1_{S \otimes X}$. Now applying the functor $\mathrm{Hom}_R(X, -)$, we get the following diagram in Mod-S,

$$\begin{array}{ccc} S & \xrightarrow{u} & L \\ \downarrow{\sigma_S} & & \downarrow{\sigma_L} \\ \mathrm{Hom}(X, S \otimes X) & \longrightarrow & \mathrm{Hom}(X, L \otimes X), \end{array}$$

in which σ_S is an isomorphism, $\sigma_L \circ u = \mathrm{Hom}(X, u \otimes X) \circ \sigma_S$, $\mathrm{Hom}(X, 1_{S \otimes X}) \circ \sigma_S = \sigma_S$ and $\mathrm{Hom}(X, h) \circ \mathrm{Hom}(X, u \otimes X) = \mathrm{Hom}(X, 1_{S \otimes X})$.

Therefore, $\sigma_S^{-1} \circ \operatorname{Hom}(X,h) \circ \sigma_L \circ u = 1_S$, and this shows that u splits. Thus, the short exact sequence $0 \to S_S \to^u L \to F \to 0$ splits, and hence, $\operatorname{End}(X)$ is a right cotorsion ring. Therefore, by Theorem 1.132, $\operatorname{End}(X)/J(\operatorname{End}(X))$ is von Neumann regular right self-injective and idempotents lift modulo $J(\operatorname{End}(X))$. $\qquad\square$

Theorem 6.7 *Let $X \in \mathcal{X}$ be an $\mathcal{X}$-strongly purely closed module and $Y \in \mathcal{X}$ be an $\mathcal{X}$-strongly pure submodule of X. If there exists an $\mathcal{X}$-strongly pure monomorphism $u : X \to Y$, then $X \cong Y$.*

Proof As $Y \in \mathcal{X}$, Y must be a direct summand of X. Thus, we can find a submodule H of X such that $X = H \oplus Y$. Now

$$X = H \oplus Y \supseteq H \oplus u(X) = H \oplus u(H) \oplus u(Y) \supseteq \ldots,$$

and thus, calling $P = \oplus_{t=0}^{\infty} u^t(H)$, we get that $X \supseteq P$. By construction, $P \cap Y = \oplus_{t=1}^{\infty} u^t(H) = u(P)$.

Let $v_{P \cap Y} : P \cap Y \to X(P \cap Y)$ be the $\mathcal{X}$-envelope of $P \cap Y$, and call $w : P \cap Y \to Y$, the inclusion. Note that Y is an $\mathcal{X}$-strongly purely closed module, since it is a direct summand of X. Thus, as w is a directed union of inclusions of direct summands of Y, it is an $\mathcal{X}$-strongly pure monomorphism. This means that there exists a $q : Y \to X(P \cap Y)$ such that $q \circ w = v_{P \cap Y}$, since $X(P \cap Y) \in \mathcal{X}$. Similarly, as $Y \in \mathcal{X}$ and $X(P \cap Y)$ is an $\mathcal{X}$-envelope, there exists an $h : X(P \cap Y) \to Y$ such that $h \circ v_{P \cap Y} = w$.

In particular, $q \circ h \circ v_{P \cap Y} = v_{P \cap Y}$. As $v_{P \cap Y}$ is an envelope, we deduce that $q \circ h$ is an isomorphism. Therefore, h is a splitting monomorphism, and $Q = \Im(h)$ is a direct summand of Y. So there exists a submodule K such that $Y = Q \oplus K$.

Now, $X = H \oplus Y = H \oplus (Q \oplus K) = (H \oplus Q) \oplus K$. Thus, $H \oplus Q \in \mathcal{X}$. Moreover, the inclusion $i : P \to H \oplus Q$ may be viewed as $i = (1_H \oplus v_{P \cap Y}) : P = H \oplus (P \cap Y) \to H \oplus X(P \cap Y) \cong H \oplus Q$. So i is an $\mathcal{X}$-strongly pure monomorphism. Now, as $Q \in \mathcal{X}$, we deduce that there exists a $\psi : H \oplus Q \to Q$ such that $\psi \circ i = h \circ v_{P \cap Y} \circ u$. Thus, $h^{-1} \circ \psi \circ i = v_{P \cap Y} \circ u$. Note that $u : P \to P \cap Y$ and $h : X(P \cap Y) \to Q$ are isomorphisms. By the same way, as $H \oplus Q \in \mathcal{X}$, there exists a $\varphi : Q \to H \oplus Q$ such that $\varphi \circ h \circ v_{P \cap Y} \circ u = i$. This gives us $\varphi \circ \psi \circ i = i$ and $\psi \circ \varphi \circ h \circ v_{P \cap Y} = v_{P \cap Y}$. On the other hand, as $v_{P \cap Y} : P \cap Y \to X(P \cap Y)$ is an $\mathcal{X}$-envelope and $H \in \mathcal{X}$, we get that $i = (1_H \oplus v_{P \cap Y}) : P = H \oplus (P \cap Y) \to H \oplus Q$ is an $\mathcal{X}$-envelope. As both i and $v_{P \cap Y}$ are envelopes, we deduce that $\varphi \circ \psi$ and $\psi \circ \varphi \circ h$ (and, thus, $\psi \circ \varphi$) are automorphisms. Therefore, both φ and ψ are isomorphisms. Finally, $\psi \oplus 1_K : X = (H \oplus Q) \oplus K \to Q \oplus K = Y$ is the desired isomorphism. Thus, $X \cong Y$. $\qquad\square$

Applying the preceding theorem to the particular cases of injective envelopes, pure-injective envelopes and cotorsion envelopes, we obtain the following.

Corollary 6.8 *Let E be a module.*

1. *If E is an injective module and E′ is an injective submodule of E such that there exists a monomorphism $u: E \to E′$, then $E \cong E′$.*
2. *If E is a pure-injective module and E′ is a pure-injective pure submodule of E such that there exists a pure monomorphism $u: E \to E′$, then $E \cong E′$.*
3. *If E is a flat cotorsion module and E′ is a pure submodule of E such that E′ is also flat cotorsion, and there exists a pure monomorphism $u: E \to E′$, then $E \cong E′$.*

Proof The preceding theorem applies to the cases of injective, pure-injective and flat cotorsion modules in view of Example 6.5. □

The next theorem addresses the Schröder–Bernstein problem for modules invariant under endomorphisms of their general envelopes.

Theorem 6.9 *Let X be an enveloping class and M, N be two X-endomorphism invariant modules with monomorphic X-envelopes $v_M: M \to X(M)$ and $v_N: N \to X(N)$, respectively. Assume that N is X-strongly purely closed and M is an X-strongly pure submodule of N. If there exists an X-strongly pure monomorphism $u: N \to M$, then $M \cong N$.*

Proof Let $w′$ be an X-strongly pure monomorphism from M to N. As $v_M: M \to X(M)$ is an X-envelope and $v_N \circ w′: M \to X(N)$ is an X-preenvelope, there exists a split monomorphism $f_1: X(M) \to X(N)$ such that $f_1 \circ v_M = v_N \circ w′$. A similar argument shows that there exists a split monomorphism $f_2: X(N) \to X(M)$ such that $f_2 \circ v_N = v_M \circ u$. Since the composition $f_2 \circ f_1: X(M) \to X(M)$ is also a split monomorphism, there exists an endomorphism $g: X(M) \to X(M)$ such that $g \circ (f_2 \circ f_1) = 1_{X(M)}$. Moreover, as M is X-endomorphism invariant, there exists a homomorphism $\delta: M \to M$ such that $v_M \circ \delta = g \circ v_M$. This gives us $v_M \circ \delta \circ u \circ w′ = g \circ v_M \circ u \circ w′ = g \circ f_2 \circ v_N \circ w′ = g \circ f_2 \circ f_1 \circ v_M = v_M$. Since v_M is a monomorphism, $\delta \circ u \circ w′$ is an automorphism. Therefore, $w′$ is a splitting monomorphism, and this yields that M is a direct summand of N. Thus, we can find a submodule H of N such that $N = H \oplus M$. Now,

$$N = H \oplus M \supseteq H \oplus u(N)$$
$$= H \oplus u(H) \oplus u(M) \supseteq \cdots \supseteq \oplus_{i=0}^{n} u^i(H) \oplus u^n(M) \supseteq \cdots .$$

Call $P = \bigoplus_{i=0}^{\infty} u^i(H) = H \oplus (\bigoplus_{i=1}^{\infty} u^i(H)) = H \oplus (P \cap M) \subseteq N$. By construction, $u(P) = P \cap M$. Let $v_{P \cap M} : P \cap M \to X(P \cap M)$ be an $\mathcal{X}$-envelope of $P \cap M$ and $w : P \cap M \to M$ be the inclusion. As w is a directed union of inclusions of direct summands of M, it is an $\mathcal{X}$-strongly pure monomorphism. As $v_{P \cap M}$ is an $\mathcal{X}$-envelope, there exists a homomorphism $h : X(P \cap M) \to X(M)$ such that $h \circ v_{P \cap M} = v_M \circ w$. And as w is an $\mathcal{X}$-strongly pure monomorphism, there exists a homomorphism $p : X(M) \to X(P \cap M)$ such that $p \circ v_M \circ w = v_{P \cap M}$. In particular, $p \circ h \circ v_{P \cap M} = v_{P \cap M}$, and since $v_{P \cap M}$ is an $\mathcal{X}$-envelope, $p \circ h = 1_{X(P \cap M)}$. On the other hand, $h \circ p$ is an endomorphism of $X(M)$. As M is $\mathcal{X}$-endomorphism-invariant, $(h \circ p)(M) \subseteq M$. This means that, $h|_{p(M)}$ is a homomorphism from $p(M)$ to M.

Now we proceed to show that $v_{p(M)} : p(M) \to X(P \cap M)$ is an $\mathcal{X}$-envelope and $p(M)$ is $\mathcal{X}$-endomorphism-invariant. Let $X' \in \mathcal{X}$ and $f : p(M) \to X'$ be a homomorphism. As v_M is an $\mathcal{X}$-envelope and $X' \in \mathcal{X}$, there exists a homomorphism $\alpha : X(M) \to X'$ such that $\alpha \circ v_M = f \circ p|_M$. Note that $v_M \circ h|_{p(M)} = h \circ v_{p(M)}$ and $p \circ v_M = v_{p(M)} \circ p|_M$, by the definitions of the homomorphisms. Therefore, we have $\alpha \circ h : X(P \cap M) \to X'$, with $(\alpha \circ h) \circ v_{p(M)} = f$. So we deduce that $v_{p(M)} : p(M) \to X(P \cap M)$ is an $\mathcal{X}$-preenvelope. Moreover, it can be shown that $v_{p(M)} : p(M) \to X(P \cap M)$ is indeed an $\mathcal{X}$-envelope. Now, let $\varphi : X(P \cap M) \to X(P \cap M)$ be an endomorphism. As $h \circ \varphi \circ p$ is an endomorphism of $X(M)$ and M is $\mathcal{X}$-endomorphism invariant, $(h \circ \varphi \circ p)(M) \subseteq M$. So we have $\varphi(p(M)) \subseteq p(M)$. Thus, $p(M)$ is $\mathcal{X}$-endomorphism-invariant.

Furthermore, we have $v_{p(M)} = p \circ h \circ v_{p(M)} = p \circ v_M \circ h|_{p(M)} = v_{p(M)} \circ p|_M \circ h|_{p(M)}$, and as $v_{p(M)}$ is a monomorphism, we get that $p|_M \circ h|_{p(M)} = 1_{p(M)}$. Therefore, $h|_{p(M)} : p(M) \to M$ is a splitting monomorphism, and $Q = \mathrm{Im}(h|_{p(M)}) = h \circ p(M)$ is a direct summand of M. So there exists a module K such that $M = Q \oplus K$. Again, $N = H \oplus M = H \oplus (Q \oplus K) = (H \oplus Q) \oplus K$, and thus, $H \oplus Q$ is an $\mathcal{X}$-endomorphism-invariant module.

Moreover, the inclusion $i : P \to H \oplus Q$ may be viewed as $i := (1_H \oplus (p|_M \circ w)) : P = H \oplus (P \cap M) \to H \oplus p(M) \cong H \oplus Q$. So i is an $\mathcal{X}$-strongly pure monomorphism. As $H \oplus Q$ is $\mathcal{X}$-endomorphism-invariant, there exists a $\psi : H \oplus Q \to Q$ such that $\psi \circ i = h|_{p(M)} \circ p|_M \circ w \circ u|_P$, where $h|_{p(M)} : p(M) \to Q$, $p|_M : M \to p(M)$ and $u|_P : P \to P \cap M$ are isomorphisms. On the other hand, as i is an $\mathcal{X}$-strongly pure monomorphism and $H \oplus Q$ is an $\mathcal{X}$-endomorphism invariant module, we get that $i : P = H \oplus (P \cap M) \to H \oplus Q$ is an $\mathcal{X}$-envelope. Similarly, there exists a homomorphism $\varphi : Q \to H \oplus Q$ such that $\varphi \circ h|_{p(M)} \circ p|_M \circ w \circ u|_P = i$. This means that

$\varphi \circ \psi \circ i = i$ and $h|p(M)^{-1} \circ \psi \circ \varphi \circ h|p(M) \circ v_{P \cap M} = v_{P \cap M}$. And, as both i and $v_{P \cap M}$ are envelopes, we deduce that $\varphi \circ \psi$ and $h|p(M)^{-1} \circ \psi \circ \varphi \circ h|p(M)$ are automorphisms. Thus, it follows that $\psi \circ \varphi$ is also an automorphism. Therefore, both φ and ψ are isomorphisms. Finally, $\psi \oplus 1_K : N = (H \oplus Q) \oplus K \to Q \oplus K = M$ is the desired isomorphism. This completes the proof. □

Applying the preceding theorem to the particular cases of injective envelopes, pure-injective envelopes and cotorsion envelopes, we get the following.

Corollary 6.10 *Let M and N be two modules.*

1. *(Bumby's Theorem) If M and N are quasi-injective modules such that there is a monomorphism from M to N and a monomorphism from N to M, then $M \cong N$.*
2. *If M and N are pure-quasi-injective modules such that there is a pure monomorphism from M to N and a pure monomorphism from N to M, then $M \cong N$.*
3. *If M and N are flat modules invariant under endomorphisms of their cotorsion envelopes such that there is a pure monomorphism from M to N and a pure monomorphism from N to M, then $M \cong N$.*

6.2 Schröder–Bernstein Problem for Automorphism-Invariant Modules

In this section, we will see that the Schröder–Bernstein problem has a positive solution for automorphism-invariant modules and pure-automorphism-invariant modules.

Theorem 6.11 *Let M, N be automorphism-invariant modules, and let $f : M \to N$ and $g : N \to M$ be monomorphisms. Then $M \cong N$.*

Proof By Corollary 6.8, we know that $E(M) \cong E(N)$. On the other hand, we have a diagram

$$M \xrightarrow{\ f\ } f(M) \xrightarrow{\ u\ } N \xrightarrow{\ g\ } M$$
$$\downarrow{\scriptstyle 1_M}$$
$$M$$

in which $u : f(M) \to N$ is the inclusion. As M is automorphism-invariant and $g \circ u \circ f$ is monic, there exists a $\varphi : M \to M$ such that $\varphi \circ g \circ u \circ f = 1_M$.

And as $f: M \to f(M)$ is an isomorphism, this means that $u: f(M) \to N$ splits, and thus, $f(M)$ is a direct summand of N. Similarly, $g(N)$ is a direct summand of M.

As $f: M \to f(M)$ and $g: N \to g(N)$ are isomorphisms, we know that $E(f(M)) \cong E(g(N))$. We proceed to show that $f(M) \cong g(N)$. Let $h: E(g(N)) \to E(f(M))$ be an isomorphism. Call $M' = h^{-1}(f(M)) \cap g(N)$ and $N' = h(g(N)) \cap f(M)$. By construction, $h|_{M'}: M' \to N'$ is an isomorphism. Moreover, as $g(N) \subseteq_e E(g(N))$, we have that $h(g(N)) \subseteq_e E(f(M))$. Similarly, $h^{-1}(f(M)) \subseteq_e E(g(N))$. Therefore, $M' \subseteq_e E(g(N))$ and $N' \subseteq_e E(f(M))$. In particular, $M' \subseteq_e g(N)$ and $N' \subseteq_e f(M)$. We have then

$$M' \xrightarrow{h|_{M'}} N' \xrightarrow{u_{N'}} f(M),$$

$$\downarrow u_{M'}$$

$$g(N)$$

where $u_{M'}$ and $u_{N'}$ are inclusions. Moreover, $g(N)$ is a submodule of M, and $f(M)$ is isomorphic to M. Therefore, $f(M)$ is automorphism-invariant, and as $u_{N'} \circ h|_{M'}$ is monic, there exists a $\psi: g(N) \to f(M)$ such that $\psi \circ u_{M'} = u_{N'} \circ h|_{M'}$. Similarly, there exists a $\varphi: f(M) \to g(N)$ such that $\varphi \circ u_{N'} = u_{M'} \circ h^{-1}|_{N'}$. Composing, we get the diagram

$$
\begin{array}{ccccc}
M' & \xrightarrow{h|_{M'}} & N' & \xrightarrow{h^{-1}|_{N'}} & M' \\
\downarrow{\scriptstyle u_{M'}} & & \downarrow{\scriptstyle u_{N'}} & & \downarrow{\scriptstyle u_{M'}} \\
g(N) & \xrightarrow{\psi} & f(M) & \xrightarrow{\varphi} & g(N).
\end{array}
$$

So $\varphi \circ \psi \circ u_{M'} = \varphi \circ u_{N'} \circ h|_{M'} = u_{M'} \circ h^{-1}|_{N'} \circ h|_{M'} = u_{M'}$. And this means that $(1_{g(N)} - \varphi \circ \psi) \circ u_{M'} = 0$. As $u_{M'}$ is monic, we deduce that $(1_{g(N)} - \varphi \circ \psi)$ has essential kernel, and thus, $(1_{g(N)} - \varphi \circ \psi) \in J(\text{End}(g(N)))$ since $g(N)$ is automorphism-invariant. Therefore, $\varphi \circ \psi$ is an isomorphism. Similarly, $\psi \circ \varphi$ is an isomorphism, and thus, $\varphi: f(M) \to g(N)$ is an isomorphism. As $M \cong f(M)$ and $N \cong g(N)$, we deduce that $M \cong N$. □

Recall that there exists a full embedding $F: \text{Mod}{-}R \to \mathcal{C}$ of $\text{Mod}{-}R$ into a locally finitely presented Grothendieck category $\mathcal{C}$ (normally called the functor category of Mod-R) with a right adjoint functor $G: \mathcal{C} \to \text{Mod-}R$ as explained in Chapter 1.

Corollary 6.12 *Let M, N be two modules invariant under automorphisms of their pure-injective envelopes, and let $f : M \to N$ and $g : N \to M$ be pure monomorphisms. Then $M \cong N$.*

Proof In this case, $F(M), F(N)$ are automorphism-invariant objects in C, and $F(f) : F(M) \to F(N)$ and $F(g) : F(N) \to F(M)$ are monomorphisms. So $F(M) \cong F(N)$, by Theorem 6.11, and consequently, $G \circ F(M) \cong G \circ F(N)$. Note that $M \cong G \circ F(M)$ and $N \cong G \circ F(N)$. Thus, we have $M \cong N$. $\square$

The next example shows that there exist flat modules that are invariant under automorphisms of their cotorsion envelopes, but they are not invariant under endomorphisms of their cotorsion envelopes nor under automorphisms of their injective or pure-injective envelopes, and therefore, the preceding results cannot be applied to these modules.

Example 6.13 Let K be a field of characteristic zero and S be the K-algebra constructed in [141, section 2]. Then S is a right Artinian ring that is not right pure-injective. As S_S is Artinian, any right S-module is cotorsion and, thus, is invariant under automorphisms of its cotorsion envelope. Assume that any direct sum of copies of S_S is invariant under automorphisms of its pure-injective envelope. As char$(K) = 0$, this means that any direct sum of copies of S_S is also invariant under endomorphisms of its pure-injective envelope, and thus, $F(S_S)$ is Σ-quasi-injective in the functor category C. But then $E(F(S_S))$ is Σ-injective, and this means that the pure-injective envelope of S_S is Σ-pure-injective. Therefore, S_S is also Σ-pure injective, as it is a pure submodule of its pure-injective envelope, a contradiction. Thus, we conclude that there exists an index set I such that $S_S^{(I)}$ is not invariant under automorphisms of its pure-injective envelope. Call $M_S = S_S^{(I)}$.

Now let R be the ring of all eventually constant sequences over $\mathbb{F}_2$, the field of two elements. It is known that R is a von Neumann regular ring, and R_R is an automorphism-invariant module that is not quasi-injective. Therefore, it cannot be invariant under endomorphisms of its cotorsion envelope, nor of its pure-injective envelope, either.

Let us consider the ring $R \times S$ and the right $R \times S$-module $R \times M$. Then:

1. $R \times M$ is flat, and it is invariant under automorphisms of its cotorsion envelope, since so are R_R and M_R.
2. $R \times M$ is not invariant under endomorphisms of its cotorsion envelope since, otherwise, so would be R_R.

3. $R \times M$ is not invariant under automorphisms of its pure-injective envelope since, otherwise, so would be M_R.

4. $R \times M$ is not invariant under automorphisms of its injective envelope since, otherwise, it would be quasi-injective, as S is an algebra over a field of characteristic zero. And this would mean that S_S would be injective, since it is a direct summand of M_S.

Notes

This chapter is entirely based on [51].

7

Automorphism-Extendable Modules

In this chapter, we study classes of modules that generalize the notions of automorphism-invariant modules and quasi-injective (that is, endomorphism-invariant) modules.

Definition 7.1 A module M is said to be an automorphism-extendable (resp., endomorphism-extendable) module if for every submodule X of M, each automorphism (resp., endomorphism) of the module X can be extended to an endomorphism of the module M.

Clearly, every endomorphism-extendable module is automorphism-extendable and every quasi-injective module is endomorphism-extendable (and, hence, automorphism-extendable). We have already seen that if M is an automorphism-invariant module, then for any submodule N of M, any automorphism of N can be extended to an automorphism of the module M. This shows, in particular, that if M is an automorphism-invariant module, then M is automorphism-extendable as well. In addition, $\mathbb{Z}$ is an example of an automorphism-extendable (in fact, endomorphism-extendable) $\mathbb{Z}$-module that is not automorphism-invariant.

We now give some examples of modules that are automorphism-extendable but not endomorphism-extendable.

Example 7.2 Let F be a field, and let A be the F-algebra with two generators x, y and the relation $xy - yx = 1$. We claim that A_A and $_A A$ are automorphism-extendable modules but not endomorphism-extendable. It is well known that A is a simple principal right (left) ideal domain, A is not a division ring, and the group of invertible elements $U(A)$ of the domain A coincides with $F \backslash 0$. In particular, $U(A)$ is contained in the center of the domain A.

Let a be a nonzero non-invertible element of the domain A. It is sufficient to prove the following two assertions.

1. For every automorphism α of the module aA_A, there exists an invertible element $u \in U(A)$ such that $\alpha(ab) = uab$ for all $b \in A$.
2. There exists an endomorphism f of the module aA, which cannot be extended to endomorphism of the module A_A.

Since $\alpha(aA) = aA$, we have that $\alpha(a) = au$ and $a = auv$ for some elements $u, v \in A$. Then $uv = 1$. Since A is a domain, $vu = 1$ and $u \in U(A) \subset F$. Then $uab = aub = \alpha(ab)$ for all $b \in A$. This shows (1).

Since A is a simple domain and a is a nonzero non-invertible element, $AaA = A \neq Aa$. Therefore, $ab \nsubseteq Aa$ for some element $b \in A$. Since A is a Noetherian domain, A has the classical division ring of fractions that contains the element a^{-1}. Then $aba^{-1}aA \subseteq aA$, and the relation $f(ac) = aba^{-1}c$ defines an endomorphism f of the submodule aA_A in A_A. We assume that f can be extended to an endomorphism φ of the module A_A. We set $d = \varphi(a)$. Then

$$ab = aba^{-1}a = f(a)a = \varphi(a)a = da \in Aa.$$

This is a contradiction. This establishes (2).

Example 7.3 Let R be the ring of all eventually constant sequences with entries in the field of two elements $\mathbb{F}_2$. Then R is a commutative regular ring. We have already seen that R_R is an automorphism-invariant module that is not quasi-injective. It is not difficult to see that if R is a von Neumann regular ring and R_R is endomorphism-extandable, then R_R is injective. Thus, it follows that the module R_R is an automorphism-extendable module that is not endomorphism-extendable.

The following is easy to see.

Lemma 7.4 *All direct summands of automorphism-extendable (resp., endomorphism-extendable) modules are automorphism-extendable (resp., endomorphism-extendable) modules.*

Definition 7.5 A module M is called a strongly automorphism-extendable module if for every submodule X of M, each automorphism of the module X can be extended to an automorphism of the module M.

The $\mathbb{Z}$-module $\mathbb{Z}$ is an example of a strongly automorphism-extendable module that is not automorphism-invariant. The additive group of rational numbers $\mathbb{Q}$ is the injective hull of the module $\mathbb{Z}_{\mathbb{Z}}$. In addition, $\mathbb{Z}_{\mathbb{Z}}$ is not an automorphism-invariant module, since $\alpha(\mathbb{Z}) \nsubseteq \mathbb{Z}$, where $\alpha: q \to q/2$ is an automorphism $\mathbb{Z}$-module $\mathbb{Q}$. Nevertheless, it is directly verified that every non-identity automorphism α of an arbitrary nonzero submodule X of the module $\mathbb{Z}_{\mathbb{Z}}$ is the multiplication by -1; therefore, α can be extended to an automorphism of the module $\mathbb{Z}_{\mathbb{Z}}$.

Obviously, every strongly automorphism-extendable module is automorphism-extendable.

7.1 General Properties of Automorphism-Extendable Modules

In this section, we will discuss the basic properties of automorphism-extendable modules.

Lemma 7.6 *Let M be a module, let E be the injective hull of M, let X be a submodule of M, let Y be an arbitrary $\cap$-complement to X in the module M, and let $X' = X \oplus Y$.*

1. *Every automorphism (resp., endomorphism) f of the module X with the use of the relation $f'(x + y) = f(x) + y$, where $x \in X$ and $y \in Y$, can be extended to an automorphism (resp., endomorphism) f' of X'. In turn, the automorphism (resp., endomorphism) f' of the module X' can be extended to an automorphism (resp., endomorphism) α of the injective module E.*
2. *Any homomorphism $g \colon X \to M$ with essential kernel K can be extended to a homomorphism $g' \colon X' \to M$ with essential kernel $K' = K \oplus Y$ with the use of the relation $g'(x + y) = g(x)$, where $x \in X$ and $y \in Y$. In turn, the homomorphism g' can be extended to an endomorphism h of the injective module E with essential kernel, and $1_E - h$ is an automorphism of the module E that coincides with the identity automorphism of the module K'.*

Proof Since X' is an essential submodule of M and M is an essential submodule of E, we have that X' is an essential submodule of E.

1. It is directly verified that f' is an automorphism (resp., endomorphism) of the essential submodule X' in E. Since the module E is injective, the automorphism (resp., endomorphism) f' of the module X' can be extended to an endomorphism α of the module E. We assume that f' is an automorphism of the module X'. Since X' is an essential submodule in E and $X \cap \operatorname{Ker}\alpha = 0$, we have that α is a monomorphism and the module $\alpha(E)$ is injective. Therefore, $\alpha(E)$ is a direct summand of E. In addition, $X' = f'(X') = \alpha(X'))$, whence the module $\alpha(E)$ contains an essential submodule X' of the module E. Therefore, $\alpha(E) = E$ and α is an automorphism.

Let f' be an endomorphism (resp., automorphism) of the module X', and let Y be a submodule in M that is maximal among submodules in M that have the zero intersection with X'. Then $X = X' \oplus Y$ is an essential submodule in M. Now we can be define an endomorphism (resp., automorphism) f of the module X such that $f(x' + y) = f(x') + y$ for any $x' \in X'$ and $y \in Y$.

2. It is directly verified that $g': X' \to M$ is a homomorphism with essential kernel $K' = K \oplus Y$. Since the module E is injective, the homomorphism g' can be extended to an endomorphism h of the module E. Since the module $\operatorname{Ker} h$ contains an essential submodule K' of the module M, we have that $\operatorname{Ker} h$ is an essential submodule in E. Then the restriction of the endomorphism $1_E - h$ of the injective module E to the module $\operatorname{Ker} h$ is the identity automorphism of the essential submodule $\operatorname{Ker} h$ of the module E. Then $1_E - h$ is an essential injective submodule of the module E. Therefore, $1_E - h$ is an automorphism of the module E that coincides with the identity automorphism of the module K' at the essential submodule K' of M.   □

From the preceding lemma, we have the following.

Corollary 7.7 *A module M is strongly automorphism-extendable if and only if for any essential submodule X of M, every automorphism of X can be extended to an automorphism of the module M. In addition, the module M is automorphism-extendable (resp., endomorphism-extendable) if and only if for every essential submodule X of M, each automorphism (resp., endomorphism) of the module X can be extended to an endomorphism of the module M.*

We have already seen that every automorphism-invariant module M satisfies the property C_2.

Theorem 7.8 *Let M be a strongly automorphism-extendable module. Then M satisfies the property C_3. In addition, if M is a CS module, then M is a quasi-continuous module.*

Proof Let $M = A \oplus A' = B \oplus B'$ and let $A \cap B = 0$. We have to prove that $A \oplus B$ is a direct summand in M. Let $\pi: M \to A'$ be the projection with kernel A. There exists a submodule C in M such that $(A + B) \cap C = 0$ and $A \oplus B \oplus C$ is an essential submodule in M. We set $D = B \oplus C$. Then $A \oplus D = A \oplus \pi D$ and $\pi|_D: D \to \pi D$ is an isomorphism. Therefore, $1_A \oplus \pi_D: A \oplus D \to A \oplus \pi D$ is an automorphism of the module $A \oplus D$. Since M is a strongly automorphism-extendable module, $1_A \oplus \pi_D$ can be extended to an automorphism f of the module M. Since B is a direct summand of the module M, $\pi B = f B$ is a direct summand of the module M. Therefore, πB is a direct summand of the module A', whence $A \oplus B = A \oplus \pi B$ is a direct summand of the module M. Therefore, M is a C_3 module.

Now the second assertion follows from the definition of a quasi-continuous module.   □

Note that $\mathbb{Z}_{\mathbb{Z}}$ is a strongly automorphism-extendable quasi-continuous module that is not automorphism-invariant.

Lemma 7.9 *If M is an automorphism-extendable right A-module and X and Y are two submodules in M with $X \cap Y = 0$, then for every $f \colon Y \to X$ homomorphism, there exists an endomorphism g of the module M that coincides with $f \colon Y \to X$ on Y.*

Proof We define an endomorphism α of the module $X \oplus Y$ by the relation $\alpha(x + y) = x + f(y) + y$ for all $x \in X$ and $y \in Y$. We assume that

$$0 = \alpha(x + y) = x + f(y) + y, \quad x \in X, \; y \in Y.$$

Then α is a monomorphism, since

$$y = -x - f(y) \in X \cap Y = 0, \quad f(y) = 0,$$
$$x = x + f(y) + y = \alpha(x + y) = 0.$$

In addition, for any $x \in X$ and $y \in Y$, we have

$$x + y = (x - f(y)) + (f(y) + y) = \alpha(x - f(y)) + \alpha(y) \in \alpha(X \oplus Y).$$

Therefore, α is an automorphism of the module $X \oplus Y$. In addition, the endomorphism $\alpha - 1_{X \oplus Y}$ of the module $X \oplus Y$ coincides with the homomorphism $f \colon Y \to X$ on Y. Since M is an automorphism-extendable module, the automorphism α of the module $X \oplus Y$ can be extended to an endomorphism β of the module M. We denote by g the endomorphism $\beta - 1_M$ of the module M. Then g coincides with f on Y. $\qquad\square$

Proposition 7.10 *If M is an automorphism-extendable right R-module and $M = X \oplus Y$, then the module X is injective with respect to Y.*

Proof Let Y_1 be a submodule in Y, and let f_1 be a homomorphism from Y_1 into X. By Lemma 7.9, there exists an endomorphism g of the module M that coincides with $f_1 \colon Y_1 \to X$ on Y_1. Let π be the projection of the module $M = X \oplus Y$ onto X with kernel Y and let $u \colon Y \to M$ be the natural embedding. We denote by f the homomorphism $\pi g u$ from Y into X. Then f coincides with f_1 on Y_1. Therefore, the module X is injective with respect to Y. $\qquad\square$

Lemma 7.11 *Let M be an automorphism-extendable module such that for any endomorphism h of M with essential kernel, the endomorphism $1_M - h$ of the module M is an automorphism. Then M is a strongly automorphism-extendable module.*

Proof Let X be a submodule of M, and let f be an automorphism of the module X. We have to prove that f can be extended to an automorphism of the module M. Without loss of generality, we can assume that X is an essential submodule of M. Since M is an automorphism-extendable module, f and f^{-1} are extended to endomorphisms α and β of the module M, respectively. We denote by h_1 and h_2 the endomorphisms $1_M - \beta\alpha$ and $1_M - \alpha\beta$ of the module M, respectively. Since $h_1(X) = 0 = h_2(X)$, we have that $\operatorname{Ker} h_1$ and $\operatorname{Ker} h_2$ are essential submodules in M. Since $\beta\alpha = 1_M - h_1$ and $\alpha\beta = 1_M - h_2$, it follows from the assumption that $\beta\alpha$ and $\alpha\beta$ are automorphisms of the module M. Therefore, α is an automorphism of the module M.   □

Lemma 7.12 *Let M be an automorphism-extendable module such that for every element $x \in M$ and each endomorphism $h \in \operatorname{End}(M)$ with essential kernel, there exists a positive integer $n = n(x,h)$ with $h^n(x) = 0$. Then M is a strongly automorphism-extendable module.*

Proof Let $h \in \operatorname{End}(M)$ and let $\operatorname{Ker} h$ be an essential submodule in M. By Lemma 7.11, it is sufficient to prove that the endomorphism $1_M - h$ of the module M is an automorphism. We construct a formal series $1_M + \sum_{k=1}^{\infty} h^k$. For every element $x \in M$, there exists a positive integer $n = n(x,h)$ with $h^n(x) = 0$. Therefore, $1_M + \sum_{k=1}^{\infty} h^k$ is a correctly defined endomorphism of the module M. It is directly verified that

$$(1_M - h)\left(1_M + \sum_{k=1}^{\infty} h^k\right) = \left(1_M + \sum_{k=1}^{\infty} h^k\right)(1_M - h) = 1_M.$$

Therefore, $1_M - h$ is an automorphism of the module M.   □

Lemma 7.13 *Let M be a module, let X be a Noetherian submodule in M, and let h be an endomorphism of the submodule M such that $\operatorname{Ker} h$ is an essential submodule in M. Then there exists a positive integer $n = n(X,h)$ with $h^n(X) = 0$.*

Proof We set $X_0 = 0$ and $X_i = X \cap \operatorname{Ker} h^i$, $i = 1,2,3,\ldots$. Then $X_{i-1} \subseteq X_i$ and $h(X_i) \subseteq h(X_{i-1})$, $i = 1,2,3,\ldots$. Since X is a Noetherian module, $X_n = X_{n+1}$ for some positive integer n. Let $f \colon X \to M$ be the restriction of the homomorphism h^n to the module X. Since $X_n = X \cap \operatorname{Ker} h_n = \operatorname{Ker} f$, the homomorphism f induces the isomorphism $g \colon X/X_n \to g(X/X_n) \subseteq M$. Since $\operatorname{Ker} h$ is an essential submodule of the module M, we have that $\operatorname{Ker} h \cap g(X/X_n)$ is an essential submodule of $g(X/X_n)$. Since $g \colon X/X_n \to g(X/X_n)$ is an isomorphism, $g^{-1}(\operatorname{Ker} h \cap g(X/X_n))$ is an essential submodule of X/X_n. We denote by Y is the complete pre-image in X of the submodule $g^{-1}(\operatorname{Ker} h \cap g(X/X_n))$ in X/X_n under the action of g. Then

$$h^{n+1}(Y) = h(h^n(Y)) = h(f(Y)) = h(g(g^{-1}(\operatorname{Ker} h \cap g(X/X_n))))$$
$$\subseteq h(\operatorname{Ker} h \cap g(X/X_n)) = 0.$$

Therefore, $Y \subseteq X_{n+1}$ and $Y/X \subseteq X_{n+1}/X_n = 0$. Then $\mathrm{Ker}\, h \cap g(X/X_n) = g(Y/X) = 0$. Since $\mathrm{Ker}\, h \cap g(X/X_n)$ is an essential submodule of $g(X/X_n)$, we have $g(X/X_n) = 0$. Therefore, $h^n(X) = f(X) = g(X/X_n) = 0$, which is required. $\qquad\square$

Theorem 7.14 *If M be an automorphism-extendable locally Noetherian module, then M is a strongly automorphism-extendable module.*

In particular, every automorphism-extendable right module over a right Noetherian ring is a strongly automorphism-extendable module.

Proof Let X be an arbitrary cyclic submodule of M, and let h be an endomorphism of the module M such that $\mathrm{Ker}\, h$ is an essential submodule of M. By Lemma 7.13, there exists a positive integer $n = n(X,h)$ with $h^n(X) = 0$. By Lemma 7.12, M is a strongly automorphism-extendable module. $\qquad\square$

Proposition 7.15 *Let M be a module and let $M = \sum_{i \in I} M_i$, where each module M_i is an essential quasi-injective submodule in M. Then M is a quasi-injective module.*

Proof Let Q be the injective hull of the module M. Since all the modules M_i are essential submodules in M, all the modules M_i are essential submodules of the injective module Q. Therefore, Q is the injective hull of the module M_i for any $i \in I$. Let f be an endomorphism of the module Q. Since all the modules M_i are quasi-injective, $f(M_i) \subseteq M_i$ for any $i \in I$. Then

$$f(M) = f\left(\sum_{i \in I} M_i\right) = \sum_{i \in I} f(M_i) \subseteq \sum_{i \in I} M_i = M.$$

Therefore, the module M is quasi-injective. $\qquad\square$

7.2 Semi-Artinian Automorphism-Extendable Modules

We first recall some basic facts about semi-Artinian modules.

Proposition 7.16 *1. Every right module over a right semi-Artinian ring is semi-Artinian, and all semiprimary rings are semi-Artinian. In particular, every module over a semiprimary ring is semi-Artinian.*

2. If M is a module such that all its cyclic submodules are Artinian, then M is a semi-Artinian module. In particular, every Artinian module is a semi-Artinian.

3. *Every torsion abelian group is a semi-Artinian $\mathbb{Z}$-module, and any direct sum of infinite number of nonzero torsion abelian groups is a non-Artinian semi-Artinian $\mathbb{Z}$-module.*

The preceding assertions are well known and are directly verified. Next, we give an example to show that a semi-Artinian ring need not be semiprimary.

Example 7.17 Let F be a field, and let R be the ring of all sequences of elements in F that stabilize eventually. Then R is a commutative semi-Artinian ring that is not semiprimary.

Theorem 7.18 *For a semi-Artinian module M, the following conditions are equivalent.*

1. *M is an automorphism-invariant module.*
2. *M is a strongly automorphism-extendable module.*
3. *M is an automorphism-extendable module.*

Proof The implication (1) $\Rightarrow$ (3) is obvious.

(3) $\Rightarrow$ (2). Let X_1 be a submodule in M, and let f_1 be an automorphism of X_1. We have to prove that f_1 can be extended to an automorphism of the module M. By Theorem 4.1(e), we can assume that X_1 is an essential submodule in M.

We denote by W the set of all pairs (X', f') such that X' is a submodule in M, $X_1 \subseteq X'$, f' is an automorphism of X' and f' coincides with f_1 on the module X_1. We define a partial order on W such that $(X', f') \leq (X'', f'')$ if and only if $X' \subseteq X''$ and f'' coincides with f' on X'. It is directly verified that the union of any ascending chain of pairs from W belongs to W. By the Zorn lemma, there exists a maximal pair (X, f). Then $X = X_2$ for any pair $(X_2, f_2) \in W$ such that $X \subseteq X_2$ and f_2 coincides with f on X.

If $X = M$, then f is an automorphism of the module M; this is required.

We assume that $X \neq M$. Then the nonzero semi-Artinian module M/X is an essential extension of its nonzero socle Y/X, where X is a proper submodule of the module $Y \subseteq M$. Since M is an automorphism-extendable module, the automorphisms f and f^{-1} of the module X can be extended to endomorphisms g and h of the module M, respectively. Then $(1 - gh)(X) = 0 = (1 - hg)(X)$. Since $X \cap \operatorname{Ker} g = X \cap \operatorname{Ker} f = 0$ and X is an essential submodule in M, we have that g is a monomorphism. In addition, the restriction g to X is an automorphism of the module X. Therefore, g induces the monomorphism $\overline{g}: M/X \to M/X$. Since the socle Y/X of the module M/X is a fully invariant submodule in M/X, we have $\overline{g}(Y/X) \subseteq Y/X$.

Therefore, $g(Y) \subseteq Y$, and $X = g(X) \subsetneq g(Y)$. We similarly obtain that $h(Y) \subseteq Y$, and $X = h(X) \subsetneq h(Y)$.

Since M is a semi-Artinian module, the nonzero socle S of M is an essential submodule in M. Then $S \subseteq X$, since X is an essential submodule in M and S is a semisimple submodule in M. In addition, $(1 - gh)(X) = (1 - ff^{-1})(X) = 0$ and Y/X is a semisimple module. Therefore, $(1 - gh)(Y) \subseteq S \subseteq X$. Then

$$Y \subseteq (1 - gh)(Y) + gh(Y) \subseteq X + g(Y) = g(Y).$$

Therefore, the restriction g_Y of the monomorphism g to Y is an automorphism of the module Y and g_Y is an extension of the automorphism f of the module X. This contradicts to the choice of X.

$(2) \Rightarrow (1)$. Let E be the injective hull of the module M, let u be an automorphism of the module E, and let S be the socle of the module E. We have to prove that $u(M) \subseteq M$.

Since M is an essential semi-Artinian submodule of E, we have that S is contained in M and S is an essential submodule in M. Since S is the socle of the module E, we have $u(S) = S = u^{-1}(S)$. We denote by X the sum of all submodules X' in M such that $u(X') = X' = u^{-1}(X')$. Then $u(X) = X = u^{-1}(X)$, $S \subseteq X$, and X is an essential submodule in M.

If $X = M$, then $u(M) = M$, which is required.

We assume that $X \neq M$. Then nonzero semi-Artinian module M/X is an essential extension of its nonzero socle Y/X, where $X \subsetneq Y \subseteq M$. Since $u(X) = X = u^{-1}(X)$ and M is a strongly automorphism-extendable module, there exist automorphisms f and g of the module M such that $(u - f)(X) = 0$ and $(u^{-1} - g)(X) = 0$. In addition, Y/X is a semisimple module. Therefore, $(u - f)(Y) \subseteq S \subseteq X$ and $(u^{-1} - g)(Y) \subseteq S \subseteq X$. Since f and g are automorphisms of the module M and $f(X) = X = g(X)$, we have that f and g induce the automorphisms $\overline{f}$ and $\overline{g}$ of the module M/X with nonzero socle Y/X. Since Y/X is the socle of the module M/X, we have that $\overline{f}(Y/X) = Y/X = \overline{g}(Y/X)$, and $f(X) = X = g(X)$. Therefore, $f(Y) = Y = g(Y)$. In addition, $(u - f)(Y) \subseteq Y$ and $(u^{-1} - g)(Y) \subseteq Y$. Then $u(Y) \subseteq (u - f)(Y) + f(Y) \subseteq Y$ and $u^{-1}(Y) \subseteq (u^{-1} - g)(Y) + g(Y) \subseteq Y$. Therefore, $u(Y) = Y = u^{-1}(Y)$ and Y strongly contains X; this contradicts to the choice of X. $\qquad\square$

Corollary 7.19 *Let R be a ring, and let M be an R-module. If M is an Artinian module or R is a semiprimary ring, then the following conditions are equivalent.*

1. M is an automorphism-invariant module.
2. M is a strongly automorphism-extendable module.
3. M is an automorphism-extendable module.

Proposition 7.20 *Let M be an automorphism-extendable module, and let $M = \oplus_{i \in I} M_i$.*

1. *If M_i is a quasi-injective module for any $i \in I$, then M is a quasi-injective module.*
2. *If M_i is an automorphism-invariant uniform module for any $i \in I$, then M is a quasi-injective module.*
3. *If M_i is a semi-Artinian uniform module for any $i \in I$, then M is a quasi-injective module.*

Proof 1. Since M is an automorphism-extendable module and $M = M_i \oplus \oplus_{j \neq i} M_j$ for any $i \in I$, it follows from Proposition 7.10 that for any $i \in I$, the module $\oplus_{j \neq i} M_j$ is M_i-injective. Therefore, M is a quasi-injective module.

2. We fix $i \in I$. By 1, it is sufficient to prove that M_i is a quasi-injective module. Let E_i be the injective envelope of the module M_i and let f be an endomorphism of the module E_i. It is sufficient to prove that $f(M_i) \subseteq M_i$.

Assume that $\text{Ker } f = 0$. Then $f(E_i)$ is a nonzero injective submodule of the indecomposable module E_i. Therefore, $f(E_i) = E_i$ and f is an automorphism of the module E_i. Since E_i is the injective envelope of the automorphism-invariant module M_i, we have $f(M_i) \subseteq M_i$.

Now we assume that $\text{Ker } f \neq 0$. Since $\text{Ker } f \cap \text{Ker}(f - 1) = 0$ and E_i is a uniform module, we have that $\text{Ker}(f - 1) = 0$, $f - 1$ is a monomorphism, and $(f - 1)(E_i)$ is a nonzero injective submodule of the indecomposable module E_i. Therefore, $(f - 1)(E_i) = Q_i$ and $f - 1$ is an automorphism of the module E_i. Since M_i is automorphism-invariant, we have $(f - 1)(M_i) \subseteq M_i$. Therefore, $f(M_i) \subseteq (f - 1)(M_i) + M_i \subseteq M_i$.

We proved that $f(M_i) \subseteq M_i$ for $\text{Ker } f = 0$ and for $\text{Ker } f \neq 0$, which is required.

3. We fix $i \in I$. Since M is an automorphism-extendable module, M_i is an automorphism-extendable module. In addition, M_i is a semi-Artinian module. By Theorem 7.18, M_i is an automorphism-invariant module. By (2), M is a quasi-injective module. □

Theorem 7.21 *If R is an Artinian serial ring and M is an R-module, then the following conditions are equivalent.*

1. *M is an automorphism-extendable module.*
2. *M is an automorphism-invariant module.*

3. *M is a strongly automorphism-invariant module.*
4. *M is a quasi-injective module.*

Proof The implications $(4) \Leftarrow (3) \Leftarrow (2) \Leftarrow (1)$ are true for modules over arbitrary rings.

$(1) \Leftarrow (4)$. We have $M = \sum_{i \in I} M_i$, where each module M_i is a uniserial module of finite length. We fix $i \in I$. Since R is an Artinian ring, M_i is a semi-Artinian module. By Proposition 7.20, M is a quasi-injective module. $\square$

Remark 7.22 If R is a ring such that every right R-module is a direct sum of uniform modules, then every automorphism-extendable right R-module M is a quasi-injective module. Indeed, R is an Artinian ring; e.g. see [31, theorem 1]. Therefore, every R-module is semi-Artinian. By Proposition 7.20, M is a quasi-injective module.

7.3 Automorphism-Extendable Modules that Are Not Singular

Lemma 7.23 *Let R be a ring, and let Y be a right R-module that is not an essential extension of a singular module. Then there exists a nonzero right ideal I of the ring R such that the module I_R is isomorphic to the submodule of the module Y.*

Proof Since the module Y is not an essential extension of a singular module, there exists an element y of the module Y such that yR is a nonzero nonsingular module. Since $yR \cong R_R/r(y)$ and the module yR is nonsingular, the right ideal $r(y)$ is not essential. Therefore, there exists a nonzero right ideal I such that $I \cap r(y) = 0$. In addition, there exists an epimorphism $f : R_R \to yR$ with kernel $r(y)$. Since $I \cap \text{Ker } f = 0$, we have that f induces the monomorphism $g : I \to yR$. Therefore, yR contains the nonzero submodule $g(I)$ that is isomorphic to the module I_R. $\square$

Proposition 7.24 *Let R be a ring, and let M be an automorphism-extendable right R-module.*

1. *If X and Y are two submodules in M with $X \cap Y = 0$ and the module M/X is nonsingular, then the module X is injective with respect to Y.*
2. *If R is a right nonsingular ring and Y is any nonsingular submodule in M, then the module $Z(M)$ is injective with respect to Y.*
3. *If the module M is nonsingular and X, Y are two closed submodules in M with $X \cap Y = 0$, then the module X is injective with respect to Y and the module Y is injective with respect to X.*

Proof 1. Let Y_1 be a submodule of the module Y, and let $f_1 : Y_1 \to X$ be a homomorphism. We have to prove that f_1 can be extended to a homomorphism $f : Y \to X$. Without loss of generality, we can assume that Y_1 is an essential submodule in Y. (Indeed, it follows from the Zorn lemma that there exists a submodule Z in Y such that $Y_1 \cap Z = 0$ and $Y_1 \oplus Z$ is an essential submodule in Y. We set $Y_2 = Y_1 \oplus Z$. The homomorphism $f_1 : Y_1 \to X$ can be extended to the homomorphism $f_2 : Y_2 \to X$ with the use of the relation $f_2(y_1 + z) = f_1(y_1)$.)

We define endomorphism α of the module $X \oplus Y_1$ by the relation $\alpha(x + y_1) = x + f(y_1) + y_1$ for all $x \in X$ and $y_1 \in Y_1$. We assume that

$$0 = \alpha(x + y_1) = x + f(y_1) + y_1, \quad x \in X, \ y_1 \in Y_1,$$

$$y_1 = -x - f(y_1) \in X \cap Y_1 = 0, \quad f(y_1) = 0,$$

$$x = x + f(y_1) + y_1 = \alpha(x + y_1) = 0.$$

In addition, for any $x \in X$ and $y_1 \in Y_1$, we have

$$x + y_1 = (x - f(y_1)) + (f(y_1) + y_1)$$

$$= \alpha(x - f(y_1)) + \alpha(y_1) \in \alpha(X \oplus Y_1).$$

Therefore, α is an automorphism of the module $X \oplus Y_1$. It directly follows from the construction of α that the endomorphism $\alpha - 1$ of the module $X \oplus Y_1$ coincides with the homomorphism $f_1 : Y_1 \to X$ at the module Y_1. Since M is an automorphism-extendable module, the automorphism α of the module $X \oplus Y_1$ can be extended to an endomorphism β of the module M. We denote by g endomorphism $\beta - 1$ of the module M. Then g coincides with f_1 at Y_1.

We prove that $g(y) \in X$ for every element y of the module Y. Let $h : M \to M/X$ be the natural epimorphism. Since Y_1 is an essential submodule in Y, it follows that $yB \subseteq Y_1$ for some essential right ideal B of the ring R. Then

$$g(y)B = g(yB) \subseteq g(Y_1) = f_1(Y_1) \subseteq X,$$

$$h(g(y))B = h(g(y)B) \subseteq h(X) = 0.$$

Therefore, $h(g(y)) \in Z(M/X) = 0$ and $g(y) \in \operatorname{Ker} h = X$.

Since $g(Y) \subseteq X$, we have that g induces the homomorphism $f : Y \to X$. Therefore, the module X is injective with respect to Y.

(2) We set $X = Z(M)$. Since R is a right nonsingular ring, it follows that the module M/X is nonsingular. Since the module X is singular and the module Y is nonsingular, it follows from (1) that the module X is injective with respect to Y.

(3) Since X and Y are closed submodules nonsingular of the module M, it follows that the modules M/X and M/Y are nonsingular. By (1), X is injective with respect to Y and the module Y is injective with respect to X. □

Proposition 7.25 *Let M be a module, and let E be the injective hull of M. The following conditions are equivalent.*

1. *M is an automorphism-extendable module, and for every essential submodule Y in M, each homomorphism $Y \to M$ with essential kernel can be extended to an endomorphism of the module M with essential kernel.*
2. *M is a strongly automorphism-extendable module, and for every submodule Y in M, each homomorphism $Y \to M$ with essential kernel can be extended to an endomorphism of the module M with essential kernel.*
3. *$\alpha(M) \subseteq M$ for every automorphism α of the module E such that $\alpha(X) = X$ for some essential submodule X of the module M.*
4. *$\alpha(M) = M$ for every automorphism α of the module E such that $\alpha(X) = X$ for some essential submodule X of the module M.*

Proof The implications $(4) \Rightarrow (3)$ and $(2) \Rightarrow (1)$ are directly verified.

$(1) \Rightarrow (3)$. Let α be an automorphism of the module E, and let $\alpha(X) = X$ for some essential submodule X of the module M. We denote by Y the submodule

$$\alpha^{-1}(M \cap \alpha(M)) = \{y \in M | \alpha(y) \in M\}$$

of the module M. Then $\alpha(Y) \subseteq M$, $X \subseteq Y$ and Y is an essential submodule of M. In addition, α induces the automorphism φ_1 of the module X. Since M is an automorphism-extendable module, φ_1 can be extended to an endomorphism φ_2 of the module M. Since the module E is injective, φ_2 can be extended to an endomorphism φ of the module E. We denote by g the restriction of the homomorphism $\alpha - \varphi$ to the module Y. Since $\varphi(Y) \subseteq M$, $\alpha(Y) \subseteq M$ and $g(X) = 0$, we have that g is a homomorphism from Y into M with essential kernel. By the assumption, g can be extended to an endomorphism g_1 of the module M. Since the module E is injective, g_1 can be extended to an endomorphism β of the module E. Then $(\alpha - \varphi - \beta)(Y) = (g - \beta)(Y) = 0$. We denote by Z the submodule $\{z \in M \mid (\alpha - \varphi - \beta)(z) \in M\}$ of the module M. Then Z is a complete pre-image in M of the module $M \cap (\alpha - \varphi - \beta)(M)$ under the action of the homomorphism $\alpha - \varphi - \beta$, $Y \subseteq Z$ and

$$\alpha(Z) \subseteq (\alpha - \varphi - \beta)(Z) + (\varphi + \beta)(Z) \subseteq M.$$

Therefore, $Y \subseteq Z \subseteq Y$ and $Z = Y$. Then

$$(\alpha - \varphi - \beta)(Z) = (\alpha - \varphi - \beta)(Y) = 0.$$

If $(\alpha - \varphi - \beta)(M) = 0$, then $\alpha(M) = (\varphi - \beta)(M) \subseteq M$, which is required.

We assume that $(\alpha - \varphi - \beta)(M) \neq 0$. Since M is an essential submodule of E, we have that $M \cap (\alpha - \varphi - \beta)(M)$ is an essential submodule of the nonzero module $(\alpha - \varphi - \beta)(M)$. Since Z is the complete pre-image in M nonzero of the module $M \cap (\alpha - \varphi - \beta)(M)$ under the action of the homomorphism $\alpha - \varphi - \beta$, we have $(\alpha - \varphi - \beta)(Z) \neq 0$. This is a contradiction.

$(3) \Rightarrow (4)$. Let X be an essential submodule of the module M, and let α be an automorphism of the module E such that $\alpha(X) = X$. It follows from (3) that $\alpha(M) \subseteq M$ and $\alpha^{-1}(M) \subseteq M$. Then $\alpha(M) = M$.

$(4) \Rightarrow (2)$. Let X be a submodule of the module M, and let φ be an automorphism of the module X. By Lemma 7.6 and the property that M is an essential submodule of E, we can assume that X is an essential submodule of E. By Lemma 7.6, the automorphism φ of the module X can be extended to an automorphism α of the injective module E. By the assumption, $\alpha(M) = M$. Therefore, φ can be extended to an automorphism of the module M.

Let $h_1 \colon Y \to M$ be a homomorphism with essential kernel K_1. By Lemma 7.6, we can assume that Y and K_1 are essential submodules in E. The homomorphism h_1 can be extended to an endomorphism h of the injective module E, and h has the essential kernel K. We denote by α the endomorphism $1_E - h$ of the module E. Then the restriction of endomorphism $1_E - h$ to K is the identity automorphism of the essential submodule K in the injective module E. Therefore, $\mathrm{Ker}(1_E - h) = 0$ and $(1_E - h)(E)$ is an injective essential submodule of E. Therefore, $1_E - h$ is an automorphism of the module E and $(1_E - h)(K) = K$. By (4), $(1_E - h)(M) = M$. Therefore, $(1_E - h)\big|_M$ is an automorphism of the module M. Then $1_M - (1_E - h)\big|_M$ is an endomorphism of the module M with essential kernel and $1_M - (1_E - h)\big|_M$ coincides with h_1 at Y. $\qquad \square$

Theorem 7.26 *Let M be a module, E be the injective hull of the module M, and let $M = T \oplus U$, where T is an injective module and U is a nonsingular module. The following conditions are equivalent.*

1. *M is an automorphism-extendable module.*
2. *M is a strongly automorphism-extendable module.*
3. *$\alpha(M) \subseteq M$ for every automorphism α of the module E such that $\alpha(X) = X$ for some essential submodule X of the module M.*

Proof The implication $(3) \Rightarrow (2)$ follows from Proposition 7.25.

The implication $(2) \Rightarrow (1)$ is obvious.

$(1) \Rightarrow (3)$. Let Y be an essential submodule in M, $h \colon Y \to M$ be a homomorphism with essential kernel, and let $\pi \colon M = T \oplus U \to U$ be the projection with kernel T. Then the module $\pi h(Y)$ is singular, and it is contained in nonsingular module U. Therefore, $\pi h(Y) = 0$. Then $h(Y) \subseteq T$ and h is a homomorphism from the module Y into the module T. Since the module T is injective, h can be extended to a homomorphism $M \to T \subseteq M$. This homomorphism is the required endomorphism of the module M that extends h. $\qquad\square$

As an easy consequence of the above theorem, we have the following.

Corollary 7.27 *Let M be a nonsingular module, and let E be the injective hull of M. The following conditions are equivalent.*

1. *M is an automorphism-extendable module.*
2. *M is a strongly automorphism-extendable module.*
3. *$\alpha(M) \subseteq M$ for every automorphism α of the module E such that $\alpha(X) = X$ for some essential submodule X of the module M.*

Theorem 7.28 *Let $M = T \oplus U$, where T is an injective module, U is a nonsingular module, and $\operatorname{Hom}(T', U) = 0$ for every submodule T' of the module T. The following conditions are equivalent.*

1. *M is an automorphism-extendable module.*
2. *M is a strongly automorphism-extendable module.*
3. *U is an automorphism-extendable module.*
4. *U is a strongly automorphism-extendable module.*

Proof The equivalence (1) $\Leftrightarrow$ (2) follows from Theorem 7.26.

The implication (1) $\Rightarrow$ (3) is directly verified.

The equivalence (3) $\Leftrightarrow$ (4) follows from Corollary 7.27.

(3) $\Rightarrow$ (1). Let E be an injective hull of the module M, let X be an essential submodule of the module M, and let α be an automorphism of E such that $\alpha(X) = X$. By Theorem 7.26, it is sufficient to prove that $\alpha(M) \subseteq M$. For the injective hull E of the module $M = T \oplus U$, there exists a direct decomposition $E = T \oplus U_1$, where U_1 is the injective hull of the nonsingular module U. Since α is an automorphism of the module $T \oplus U_1$ and $\operatorname{Hom}(T', U) = 0$ for every submodule T' in the module T, it is directly verified that $\alpha(T) = T$. Let $h: E \to E/T$ be the natural epimorphism. (We can assume that h is the projection from the module $E = T \oplus U_1$ onto the module U_1 with kernel T.) Then α induces the automorphism α_1 of the injective hull $h(E)$ of the module $h(U)$. Since $\alpha(X) = X$, we have $\alpha_1(h(X)) = h(X)$. By applying Corollary 7.27 to the automorphism-extendable nonsingular module $h(M) = h(U)$, we obtain that $\alpha_1(h(M)) \subseteq h(M)$. Then $\alpha(M) \subseteq M + T = M$. $\qquad\square$

Corollary 7.29 *Let $M = T \oplus U$, where T is an injective module that is an essential extension of a singular module, and let U be a nonsingular module. The following conditions are equivalent.*

1. *M is an automorphism-extendable module.*
2. *M is a strongly automorphism-extendable module.*
3. *U is an automorphism-extendable module.*
4. *U is a strongly automorphism-extendable module.*

Proposition 7.30 *Let R be a ring, let M be an automorphism-extendable right R-module, and let X, Y be two submodules of M with $X \cap Y = 0$. If the module M/X is nonsingular, then the module X is injective with respect to Y.*

Proof Let Y_1 be a submodule in Y, and let f_1 be a homomorphism from Y_1 into X. By the Zorn lemma, there exists a submodule Z in Y such that $Y_1 \cap Z = 0$ and $Y_1 \oplus Z$ is an essential submodule in Y. We set $Y_2 = Y_1 \oplus Z$. The homomorphism $f_1 \colon Y_1 \to X$ can be extended to a homomorphism $f_2 \colon Y_2 \to X$ with the use of the relation $f_2(y_1 + z) = f_1(y_1)$. By Lemma 7.9, there exists an endomorphism g of the module M that coincides with $f_2 \colon Y_2 \to X$ on Y_2.

It remains to prove that $g(y) \in X$ for any element y of the module Y. Let $h \colon M \to M/X$ be the natural epimorphism. Since Y_2 is an essential submodule in Y, the module Y/Y_2 is singular. Therefore, $yB \subseteq Y_2$ for some essential right ideal B of the ring R. Then

$$g(y)B = g(yB) \subseteq g(Y_2) = f_2(Y_2) = f_1(Y_1) \subseteq X,$$

$$h(g(y))B = h(g(y)B) \subseteq h(X) = 0.$$

Therefore, $h(g(y)) \in Z(M/X) = 0$ and $g(y) \in \operatorname{Ker} h = X$. $\square$

Lemma 7.31 *Let R be a ring, and let Y be a nonsingular right R-module. If $\{y_i\}_{i \in Y}$, $|I| \geq 2$, is a subset of the module Y such that $\operatorname{Hom}(Y_i, Y_j) = 0$ for any submodules $Y_i \subseteq y_i R$ and $Y_j \subseteq y_j R$ for all $i \neq j$, then there exists a set $\{B_i\}_{i \in I}$ of nonzero right ideals of the ring R such that for any i, the right R-module B_i is isomorphic to a submodule of $y_i R$ and $B_i B_j = 0$ for all $i \neq j$.*

Proof In I, we fix distinct subscripts i and j. By Lemma 7.23, there exist nonzero right ideals B_i and B_j of the ring R such that $y_i R$ contains a nonzero submodule Y_i that is isomorphic to the right R-module B_i, and $y_j R$ contains a nonzero submodule Y_j that is isomorphic to the right R-module B_j. By the assumption, $\operatorname{Hom}(Y_i, Y_j) = 0$. Then $\operatorname{Hom}(B_i, B_j) = 0$, since $B_i \cong Y_i$ and $B_j \cong Y_j$. For each element $b_j \in B_j$, the relation $x \to b_j x$, $x \in B_i$

defines a homomorphism from B_i into B_j. Since $\text{Hom}(B_i, B_j) = 0, b_j B_i = 0$, whence $B_i B_j = 0$. □

Proposition 7.32 *Let R be a ring, and let Y be a nonsingular square-free automorphism-right invariant R-module.*

1. *If Y is an essential extension of direct sum of uniform modules, then Y is an essential extension of the quasi-injective module M that is the direct sum of uniform quasi-injective modules.*
2. *If Y is a finite-dimensional module, then Y is an essential extension of a quasi-injective module that is the finite direct sum of uniform quasi-injective modules.*
3. *If the module Y is not finite-dimensional, then there exists an infinite set $\{B_i\}_{i=1}^{\infty}$ of nonzero right ideals of the ring R such that $B_i B_j = 0$ for all $i \neq j$.*
4. *If R is a finite subdirect product of prime rings, then Y is a finite-dimensional module that is an essential extension of the quasi-injective module $Y_1 \oplus \cdots \oplus Y_n$, where all Y_i are quasi-injective uniform modules.*

Proof 1. By the assumption, Y is an essential extension of the direct sum of uniform modules $Y_i, i \in I$. Every uniform module Y_i is an essential submodule of some closed uniform submodule M_i of the module M. We set $M = \sum_{i \in I} M_i$. Then $M = \oplus_{i \in I} M_i$ and Y is an essential extension of the module M. By Theorem 3.6, M is an automorphism-invariant module. In addition, M is the direct sum of uniform modules. By Proposition 7.20, the module M is quasi-injective. All uniform direct summands M_i of the quasi-injective module M also are quasi-injective.

2. The finite-dimensional module Y is an essential extension of the finite direct sum of uniform modules. By (1), Y is an essential extension of a quasi-injective module that is of a finite direct sum of uniform quasi-injective modules.

3. Since the module Y is not finite-dimensional, Y contains the infinite direct sum $\oplus_{i=1}^{\infty} y_i R$ of nonzero cyclic submodules. For any two distinct positive integers i, j and arbitrary submodules $Y_i \subseteq y_i R$, $Y_j \subseteq y_j R$, we have $Y_i \cap Y_j \subseteq y_i R \cap y_j R = 0$, whence $\text{Hom}(Y_i, Y_j) = 0$ by Theorem 3.6. By Lemma 7.31, there exists a set $\{B_i\}_{i=1}^{\infty}$ of nonzero right ideals of the ring R such that for any positive integer i, the right R-module B_i is isomorphic to a submodule in $y_i R$ and $B_i B_j = 0$ for all $i \neq j$.

4. By (2), it is sufficient to prove that Y is a finite-dimensional module. We assume the contrary. By (3), there exists an infinite set $\{B_i\}_{i=1}^{\infty}$ of nonzero right ideals of the ring R such that $B_i B_j = 0$ for all $i \neq j$. Since R is a finite subdirect product of prime rings, there exists a finite set $\{P_k\}_{k=1}^{n}$ of prime

ideals of the ring R with $P_1 \cap \cdots \cap P_n = 0$. We have that $B_i \neq 0$ for all i and $P_1 \cap \cdots \cap P_n = 0$. Therefore, for any positive integer i, there exists a prime ideal $P_{\alpha(i)} \in \{P_i\}_{i=1}^n$ such that B_i is not contained in $P_{\alpha(i)}$. Since $P_{\alpha(i)}$ is a prime ideal and $B_i B_j = 0 \subseteq P_{\alpha(i)}$ for all $j \neq i$, B_j is contained in $P_{\alpha(i)}$ for all $j \neq i$. In addition, B_j is not contained in $P_{\alpha(j)}$. Therefore, all ideals $P_{\alpha(i)}$ are distinct. This contradicts to the property that $\{P_k\}_{k=1}^n$ is a finite set. $\qquad\Box$

7.4 Modules over Strongly Prime Rings

Definition 7.33 A ring R is said to be a right strongly prime ring if every nonzero ideal of R contains a finite subset with zero right annihilator.

Proposition 7.34 *Let R be a ring.*

1. *If R is a domain or a simple ring, then R is a (right and left) strongly prime ring.*
2. *If R is a prime ring with the maximum condition on left annihilators, then R is a right strongly prime ring. In particular, every prime right or left Goldie ring is a right strongly prime ring.*
3. *If R is a right strongly prime ring, then R is a right nonsingular prime ring.*

Proof 1. The assertion is obvious.

2. The assertion follows from Theorem 1.1.

3. It follows from the definition of right strongly prime rings that R is a prime ring, and the ideal $Z(R_R)$ contains a finite subset $X = \{x_1, \ldots, x_n\}$ such that $r(X) = r(x_1) \cap \cdots \cap r(x_n) = 0$. All right ideals $r(x_i)$ are essential. Therefore, $r(X)$ is an essential right ideal. Since $r(X) = 0$, this is a contradiction. $\qquad\Box$

There exists a strongly prime ring R that is not right or left finite-dimensional as the following example shows. In particular, R is not a right or left Goldie ring.

Example 7.35 Let R be a free algebra in two variables over a field. Then R is a domain. In particular, R is a right and left strongly prime ring with the maximum condition on right annihilators and with the maximum on left annihilators. However, R is not a right or left finite-dimensional ring.

There exists a prime ring that is not a right strongly prime ring. This follows from Proposition 7.34 and the property that there exist prime rings that are not right nonsingular; e.g. see [78].

Proposition 7.36 *Let R be a ring, let B be a right ideal of R, let RB be the ideal generated by the right ideal B, and let X be a B_R-injective R-module.*

1. *X is an $(RB)_R$-injective module.*
2. *If the ideal RB contains a finite subset $\{y_1, \dots, y_n\}$ such that $r(\{y_1, \dots, y_n\}) = 0$, then the module X is injective.*

Proof 1. Let $\{a_i\}_{i \in I}$ be the set of all elements of the ring R. For every $i \in I$, we denote by f_i the homomorphism from B into Y_i that is defined by the relation $f_i(b) = a_i b$. Then $RB = \sum_{i \in I} f_i(B)$. The module $(RB)_R$ is a homomorphic image of the external direct sum of the modules $f_i(B, i \in I$. By Theorem 1.49, X is a $(RB)_R$-injective module.

2. Since $r(y_1) \cap \dots \cap r(y_n) = r(\{y_1, \dots, y_n\}) = 0$ and $y_i R \cong R_R / r(y_i)$ for every i, there exists a monomorphism $R_R \to \oplus_{i=1}^{n} y_i R$. By Theorem 1.50, the module RB is injective. □

Theorem 7.37 *Let R be a right strongly prime ring, and let X be a right R-module.*

1. *If X is injective with respect to some nonzero right ideal of the ring R, then X is an injective module.*
2. *If X is injective with respect to some right R-module Y that is not an essential extension of a singular module, then X is an injective module.*

Proof 1. Since the module X is injective with respect to some nonzero right ideal B, it follows from Proposition 7.36(1) that the module X is injective with respect to some nonzero ideal RB. Since the ring R is right strongly prime, the ideal RB contains a finite subset $\{y_1, \dots, y_n\}$ such that $r(\{y_1, \dots, y_n\}) = 0$. By Proposition 7.36(2), the module X is injective.

2. By Lemma 7.23, there exists a nonzero right ideal B of the ring R such that the module B_R is isomorphic to a submodule of the module Y. Since the module X is injective with respect to the module Y, it follows from Theorem 1.50 that the module X is injective. □

Proposition 7.38 *Let R be a right strongly prime ring, and let M be a right R-module that is not singular. The following conditions are equivalent.*

1. *M is an automorphism-extendable module.*
2. *$M = X \oplus Y$, where X is an injective singular module, Y is a nonzero nonsingular automorphism-extendable module, and either the module Y is uniform or Y is an injective nonuniform module.*

Proof The implication $(2) \Rightarrow (1)$ is directly verified.

$(1) \Rightarrow (2)$. We set $X = Z(M)$. By Lemma 7.34, the right strongly prime ring R is right nonsingular. By Theorem 1.16, the module M/X is nonsingular. Since the ring R is right nonsingular and the module M_R is not singular, M is not an essential extension of the singular module X. Therefore, $X \cap Y' = 0$ for some nonsingular submodule Y' in M. By Proposition 7.24(2), the module X is injective with respect to Y'. By (2), X is an injective module. Therefore, $M = X \oplus Y$, where X is an injective singular module and Y is a nonzero nonsingular automorphism-extendable module. If the module Y is uniform, then the proof is completed.

We assume that Y is a nonuniform module. Then there exist two nonzero closed submodules Y_1 and Y_2 in Y such that $Y_1 \cap Y_2 = 0$ and $Y_1 \oplus Y_2$ is an essential submodule in Y. Since Y_1 and Y_2 are closed submodules in the nonsingular module Y, the modules M/Y_1 and M/Y_2 are nonsingular. By Proposition 7.24(1), the module Y_1 is injective with respect to the nonzero nonsingular module Y_2, and the module Y_2 is injective with respect to the nonzero nonsingular module Y_1. By (2), the modules Y_1 and Y_2 are injective. Then Y is an essential extension of the injective module $Y_1 \oplus Y_2$. Therefore, $Y = Y_1 \oplus Y_2$, Y is an injective nonuniform module. $\qquad \square$

We have seen that every nonsingular automorphism-invariant module over a prime right Goldie ring is injective. In the similar spirit, we have the following.

Theorem 7.39 *Let R be a right strongly prime ring, and let M be a right R-module. The following conditions are equivalent.*

1. *M is an automorphism-invariant module.*
2. *Either M is a singular automorphism-invariant module or M is an injective module.*

Proof The implication $(2) \Leftarrow (1)$ is obvious.

$(1) \Leftarrow (2)$. The automorphism-invariant module M is an automorphism-extendable module. It follows from Proposition 7.38 that $M = X \oplus Y$, where X is an injective singular module and Y is a nonzero nonsingular automorphism-invariant module and either Y is an injective nonuniform module or the module Y is uniform. It is sufficient to consider the case, where Y is a nonzero automorphism-invariant uniform module. By Lemma 4.20, Y is a quasi-injective module. By Theorem 7.37(b), the module Y is injective. Then the module M is injective. $\qquad \square$

Theorem 7.40 *Let R be a ring, and let $G(R_R)$ be the right Goldie radical of R. Every nonsingular quasi-injective right R-module is injective if and only if $R/G(R_R)$ is a right strongly semiprime ring.*

Proposition 7.41 *Let R be a ring, and let Y be a nonsingular square-free automorphism-right invariant R-module.*

1. *If the ring R is right strongly semiprime, then Y is an injective module.*
2. *If the factor ring $R/G(R_R)$ is right strongly semiprime, then Y is an injective module.*

Proof 1. Since the ring R is right strongly semiprime, R is a finite subdirect product of prime rings, see [55, theorem 1]. By Proposition 7.32(4), Y is an essential extension of some quasi-injective module Y'. By Theorem 7.40, all nonsingular quasi-right self-injective R-modules are injective. Therefore, Y' is an injective an essential submodule of the module Y. Then $Y = Y'$ and the module Y is injective.

2. Since the module Y is nonsingular, $G(Y) = 0$. Then $YG(R_R) \subseteq G(M) = 0$. Therefore, Y is a natural right $R/G(R_R)$-module. It is directly verified that Y is a nonsingular square-free $R/G(R_R)$-module. With the use of Theorem 4.11, it is verified that Y is an automorphism-invariant $R/G(R_R)$-module. Since the factor ring $R/G(R_R)$ is right strongly semiprime, Y is an injective $R/G(R_R)$-module by (1). Therefore, Y is a quasi-injective R-module. By Theorem 7.40, Y is an injective R-module. $\square$

Corollary 7.42 *Let R be a ring, and let M be a nonsingular automorphism-invariant right R-module. If the factor ring $R/G(R_R)$ is right strongly semiprime, then M is an injective module.*

Proof By Theorem 3.6, there exists a direct decomposition $M = X \oplus Y$ such that X is a quasi-injective module, Y is a square-free nonsingular automorphism-invariant module. By Proposition 7.41(b), Y is an injective module. By Theorem 7.40, the module X is injective. Then $M = X \oplus Y$ is an injective module. $\square$

Theorem 7.43 *Let R be a ring, and let $G(R_R)$ be the right Goldie radical of R. The following conditions are equivalent.*

1. *All nonsingular automorphism-invariant right R-modules are injective.*
2. *$R/G(R_R)$ is a right strongly semiprime ring.*

Proof The implication (2) $\Rightarrow$ (1) follows from Corollary 7.42. The implication (1) $\Rightarrow$ (2) follows from the property that every quasi-injective module is an automorphism-invariant. $\qquad\square$

7.5 Modules over Hereditary Prime Rings

Proposition 7.44 *If R is a hereditary Noetherian prime ring, and M is an R-module with $r(M) \neq 0$, then the following conditions are equivalent.*

1. *M is an automorphism-extendable module.*
2. *M is an automorphism-invariant module.*
3. *M is a strongly automorphism-invariant module.*
4. *M is a quasi-injective module.*

Proof The implications (4) $\Leftarrow$ (3) $\Leftarrow$ (2) $\Leftarrow$ (1) are true for modules over arbitrary rings.

(1) $\Leftarrow$ (4). By [26, theorem 3.1], the factor ring $R/r(M)$ is an Artinian serial ring. Since M is an automorphism-extendable R-module, M is an automorphism-extendable $R/r(M)$-module. By Theorem 7.18, M is a quasi-injective $R/r(M)$-module. Therefore, M is a quasi-injective R-module. $\qquad\square$

Corollary 7.45 *Let R be a hereditary Noetherian prime ring, and let M be an automorphism-extendable R-module.*

1. *If B is a nonzero ideal of the ring R and $X = \{m \in M \,|\, mB = 0\}$, then X is a quasi-injective module.*
2. *Let $\{B_i\}_{i \in I}$ be some set of nonzero ideals of the ring R and let $X_i = \{m \in M \mid mB_i = 0\}$, $i \in I$. If $M = \sum_{i \in} X_i$ and all the modules X_i are essential submodules in M, then M is a quasi-injective module.*

Proof 1. Let f be an the endomorphism of the module M. Then

$$f(X)B = f(XB) = f(0) = 0, \quad f(X) \subseteq X.$$

Therefore, X is a fully invariant submodule of the automorphism-extendable module M. Thus, X is an automorphism-extendable R-module with nonzero annihilator. By Proposition 7.44, X is a quasi-injective module.

(2) By (1), each module M_i is quasi-injective. By Proposition 7.15, the module M is quasi-injective. $\qquad\square$

We have seen that every torsion automorphism-invariant module over a bounded hereditary Noetherian prime ring is quasi-injective. In connection to this, we have the following theorem.

Theorem 7.46 *Let R be a bounded hereditary Noetherian prime ring, and let M be a right R-module. The following conditions are equivalent.*

1. *M is an automorphism-extendable module.*
2. *M is a strongly automorphism-extendable module.*
3. *Either M is a quasi-injective singular module, or M is an injective module that is not singular, or $M = X \oplus Y$, where X is an injective singular module and Y is a nonzero automorphism-extendable uniform nonsingular module.*

Proof The implication (3) $\Leftarrow$ (2) follows from Proposition 7.38 and the property that every quasi-injective module is strongly automorphism-extendable.

The implication (2) $\Leftarrow$ (1) is true for modules over arbitrary rings.

(1) $\Leftarrow$ (3). If the module M is not singular, then it follows from Proposition 7.38 that $M = X \oplus Y$, where X is an injective singular module, Y is a nonzero nonsingular automorphism-extendable module, and the module Y is either uniform or an injective nonuniform module.

Now we assume that M is a singular automorphism-extendable module. Let $\{B_j\}_{j \in J}$ be the set of all proper invertible ideals of the ring R, and let $\{P_i\}_{i \in I}$ be the set of all maximal elements of the set $\{B_j\}_{j \in J}$. For any $i \in I$, we denote by M_i the submodule in M consisting of all elements in M that are annihilated by some power of the ideal P_i.

Any singular module M satisfies the following two properties (see [91]):

1. For any submodule X in M, we have that $X = \oplus_{i \in I} X_i$, where $X_i = X \cap M_i$, $i \in I$ and $\operatorname{Hom}(X_i, X_j) = 0$ for any distinct subscripts $i, j \in I$.
2. $M_{i,k} \subseteq M_{i,k+1}$ for every $k \in \mathbb{N}$, $M_i = \cup_{k \in \mathbb{N}} M_{i,k} = \sum_{k \in \mathbb{N}} M_{i,k}$, and $M_{i,k}$ is an essential fully invariant submodule in M_i for every $k \in \mathbb{N}$.

With the use of (1), the following two properties are directly verified.

(1) The module M is automorphism-extendable if and only if the modules M_i are automorphism-extendable.
(2) The module M is quasi-injective if and only if the modules M_i are quasi-injective.

It follows from (1) that all M_i are automorphism-extendable modules. By (2), it is sufficient to prove that all M_i are quasi-injective modules.

Then we fix $i \in I$ and denote by $M_{i,k}$, $k \in \mathbb{N}$, the submodule in M_i which is annihilated by the ideal P_i^k. By (2), $M_{i,k} \subseteq M_{i,k+1}$ for every $k \in \mathbb{N}$, $M_i = \cup_{k \in \mathbb{N}} M_{i,k} = \sum_{k \in \mathbb{N}} M_{i,k}$ and $M_{i,k}$ is an essential fully invariant submodule in M_i for every $k \in \mathbb{N}$. The fully invariant submodules $M_{i,k}$ of the

automorphism-extendable module M_i are automorphism-extendable modules
with nonzero annihilators. By Proposition 7.44, every essential submodule
$M_{i,k}$ in M_i is a quasi-injective module. By Proposition 7.15, M is a quasi-
injective module.   □

Theorem 7.47 *Let R be a bounded hereditary Noetherian prime ring and let
M be a right R-module. The following conditions are equivalent.*

1. *M is an automorphism-invariant module.*
2. *M is a quasi-injective module.*
3. *Either M is a quasi-injective singular module, or M is an injective module
   that is not singular.*

Proof The implications $(3) \Leftarrow (2) \Leftarrow (1)$ are true for modules over arbitrary
rings.

$(1) \Leftarrow (3)$. By Theorem 7.46, either M is a quasi-injective singular module,
or M is an injective module that is not singular, or $M = X \oplus Y$, where X is an
injective singular module and Y is a nonzero automorphism-invariant uniform
nonsingular module.

It is sufficient to consider only the case, where $M = X \oplus Y$, X is an
injective singular module, and Y is a nonzero automorphism-invariant uniform
nonsingular module. By Lemma 4.20, Y is a quasi-injective module. By
Theorem 7.37(2), Y is an injective module. Since $M = X \oplus Y$, the module M
is injective.   □

Remark 7.48 Let R be a bounded hereditary prime ring. The quasi-
injective R-modules are described in [90]. Therefore, Theorem 7.47 completely describes all automorphism-invariant R-modules and Theorem 7.46
describes all automorphism-extendable R-modules up to the description of
automorphism-extendable uniform nonsingular modules.

Remark 7.49 Let R be a right Ore domain, let Q be the classical right division
ring of fractions of R, and let Y be a nonzero right R-module.

1. Q_R is the injective hull of the module R_R and for every endomorphism f
   of the module Q_R, there exists an element $q \in Q$ such that $f(x) = qx$ for
   all $x \in Q_i$.
2. The module Y is a uniform nonsingular module if and only if Y is
   isomorphic to a submodule of the module Q_R. In the last case, Q_R is the
   injective hull of the module Y.
3. For any nonzero elements $q_1, \ldots, q_n \in Q$, there exist nonzero elements
   $s, a_1, \ldots, a_n \in R$ such that $q_i = a_i s^{-1}$, $i = 1, \ldots, n$. Consequently, the
   mapping

$$h: \sum_{i=1}^{n} Rq_i \to \sum_{i=1}^{n} Rq_i s \subseteq_R R$$

is a monomorphism of left R-modules.

Each assertion in Remark 7.49 is well known.

Proposition 7.50 *Let R be a right Ore domain, let Q be the classical right division ring of fractions of R, and let M be a right R-module that is not singular. The following conditions are equivalent.*

1. *M is an automorphism-extendable module.*
2. *M is a strongly automorphism-extendable module.*
3. *Either M is an injective module that is not singular, or $M = X \oplus Y$, where X is an injective singular module and Y is an automorphism-extendable module which is isomorphic to a nonzero submodule in Q_R.*

The preceding proposition follows from Remark 7.49(2) and Proposition 7.38.

Definition 7.51 A module M is said to be strongly endomorphism-extendable or completely integrally closed if for any submodule X of M, every homomorphism $X \to M$, which maps from some essential submodule in X into itself, can be extended to a homomorphism $M \to M$.

Lemma 7.52 *Let M be a module.*

1. *If M is a strongly endomorphism-extendable module, then M is endomorphism-extendable.*
2. *If M is a quasi-injective module, then M is strongly endomorphism-extendable.*

$\mathbb{Z}$ is a strongly endomorphism-extendable non-quasi-injective $\mathbb{Z}$-module.

Lemma 7.53 *Let R be an invariant hereditary domain, and let Q be the division ring of fractions of the domain R.*

1. *For any nonzero ideal B of the ring R, the factor ring R/B is the finite direct product of invariant Artinian uniserial rings.*
2. *Let M be an arbitrary submodule of any cyclic singular R-module. Then $M = M_1 \oplus \cdots \oplus M_n$, where each module M_i is a uniserial module of finite length. In addition, $f(m) \in mR$ for any element $m \in M$ and each homomorphism $f: mR \to M$.*
3. *If X is an any nonzero submodule in Q_R and M is a submodule in Q_R such that $X \subseteq M$ and M/X is a finitely generated module, then*

$\bar{f}(\bar{m}) \in \bar{m}R$ *for any element* $\bar{m} \in M/X$ *and each homomorphism*
$\bar{f} \colon \bar{m}R \to M/X$.

4. *If X is any nonzero submodule in* Q_R *and Y is a submodule in* Q_R *with*
   $X \subseteq Y$, *then* $\bar{f}(\bar{y}) \in \bar{y}R$ *for any element* $\bar{y} \in Y/X$ *and each*
   homomorphism $\bar{f} \colon \bar{y}R \to Y/X$.

5. *Every submodule Y of the module* Q_R *is a strongly automorphism-*
   extendable, strongly endomorphism-extendable module.

6. *Every uniform nonsingular R-module is a strongly automorphism-*
   extendable, strongly endomorphism-extendable module.

Proof 1. The assertion follows from [26, theorem 3.1].

2. The assertion follows from (1).

3. Since M/X is a finitely generated module, there exists a finitely
generated submodule N in M with $M = N + X$. There exists a natural
isomorphism $g \colon M/X \to N/(N \cap X)$. Since N is a finitely generated
R-submodule of the division ring of fractions Q of the domain R, it follows
from the left-right symmetric analog of Remark 7.49(3) that there exists a
monomorphism $h \colon N \to R_R$. Then the module $h(N)/h(N \cap X)$ is isomorphic
to a submodule of the cyclic singular module $R_R/h(N \cap X)$. By (2), we
have that $f(m) \in mR$ for any element $m \in h(N)/h(N \cap X)$ and each
homomorphism $f \colon mR \to h(N)/h(N \cap X)$. Since there exists a natural
isomorphism $M/X \to N/(N \cap X)$, we have that $\bar{f}(\bar{m}) \in \bar{m}R$ for any element
$\bar{m} \in M/X$ and each homomorphism $\bar{f} \colon \bar{m}R \to M/X$.

(4) Let $\bar{y} = y + X \in Y/X$, where $y \in Y$. We set $M = X + yR$. Then M/X is
a cyclic module. By (3), $\bar{f}(\bar{y}) \in \bar{y}R$ for every homomorphism $\bar{f} \colon \bar{y}R \to Y/X$.

(5) We prove that Y is a strongly endomorphism-extendable module. Let
M be a nonzero submodule in Y, let X be an essential submodule in M,
and let $g \colon M \colon Y$ be a homomorphism with $g(X) \subseteq X$. Since the module
Q_R is injective and $g(X) \subseteq X$, the homomorphism g can be extended to an
endomorphism f of the module Q_R; in addition, $f(X) \subseteq X$. Then f induces
the endomorphism $\bar{f}$ of the module Q/X. By (4), $\bar{f}(Y/X) \subseteq Y/X$. Therefore,
Y is a strongly endomorphism-extendable module.

It is similarly proved that Y is a strongly automorphism-extendable module.

(6) The assertion follows from (5) and Remark 7.49(2). □

Theorem 7.54 *Let R be an invariant hereditary domain, let Q be the division*
ring of fractions of the domain R, and let M be a right R-module. The
following conditions are equivalent.

1. *M is an automorphism-extendable module.*

2. *M is a strongly automorphism-extendable module.*

3. *M is an endomorphism-extendable module.*
4. *M is a strongly endomorphism-extendable module.*
5. *M is a quasi-injective singular module, or M is an injective module that is not singular, or $M = X \oplus Y$, where X is an injective singular module and the module Y is isomorphic to a nonzero submodule in Q_R.*

Proof The implications (5) $\Rightarrow$ (4) and (5) $\Rightarrow$ (2) follow from Lemma 7.53(5).

The implications (4) $\Rightarrow$ (3) $\Rightarrow$ (1) and (2) $\Rightarrow$ (1) are true for modules over any ring.

The implication (1) $\Rightarrow$ (5) follows from Theorem 7.46 and Remark 7.49(2).

$\square$

Proposition 7.55 *Let R be a principal right ideal domain and let U be the group of invertible elements of R. The following conditions are equivalent.*

1. *R_R is an automorphism-extendable module.*
2. *M is a strongly automorphism-extendable module.*
3. *$aU \subseteq Ua$ for any element $a \in R$.*

Proof The implication (2) $\Rightarrow$ (1) is directly verified.

(1) $\Rightarrow$ (3). We have to prove that $au \in Ua$ for any elements $a \in R$ and $u \in U$. Without loss of generality, we can assume that $a \neq 0$. We denote by φ the mapping from aR into R such that $\varphi(ab) = aub$ for any element $b \in R$. Since $aR = auR$ and R is a domain, φ is an automorphism of the module aR. Since R_R is an automorphism-extendable module, the automorphism φ can be extended to some endomorphism f of the module R_R. We set $v = f(1) \in R$. Since $vR = R$ and R is a domain, $v \in U$. Then $va = f(a) = au$ and $aU \subseteq Ua$.

(3) $\Rightarrow$ (2). Let X be a submodule of R_R, and let φ be an automorphism of X. Since R is a principal right ideal domain, $X = aR$ for some element $a \in X$. Without loss of generality, we can assume that $a \neq 0$. Since $aR = \varphi(a)R$, there exist elements $u, w \in R$ such that $\varphi(a) = au$ and $a = auw$. Since R is a domain, $1 = uw$ and $u \in U$. By the assumption, $aU \subseteq Ua$. Therefore, $au = va$ for some $v \in U$. We denote by f the automorphism of the module R_R such that $f(b) = vb$ for all $b \in R$. Then $f(a) = va = au = \varphi(a)$. Therefore, by the automorphism f is an extension of the automorphism φ. $\square$

Proposition 7.56 *Let D be a noncommutative division ring. Then $D[x]$ is a principal right (left) ideal domain that is not automorphism-extendable right or left $D[x]$-module. In addition, if the division ring D is finite-dimensional over its center F, then $D[x]$ is a bounded hereditary Noetherian prime ring.*

Proof It is well known that $D[x]$ is a principal right (left) ideal domain, and the group of invertible elements U of $D[x]$ coincides with the multiplicative group of the division ring D. In particular, $D[x]$ is a hereditary Noetherian prime ring. We assume that $D[x]_{D[x]}$ is an automorphism-extendable module. By the assumption, we have that $dd_1 \neq d_1 d$ for some nonzero elements d and d_1 of the division ring D. By Proposition 7.55, $(d + x)d_1 \subseteq Ud$. In addition, $U = D \setminus 0$. Therefore, $(d + x)d_1 = d_2(d + x)$ for some element d_2 of the division ring D. Then $d_1 x = d_2 x$ and $dd_1 = d_2 d$. Therefore, $dd_1 = d_1 d$. This is a contradiction. It can be similarly proved that the module $_{D[x]}D[x]$ also is not automorphism-extendable.

We assume that D is finite-dimensional over its center F. It is well known that for any polynomial $f \in D[x]$, there exists a polynomial $g \in D[x]$ such that fg is a nonzero polynomial in $F[x]$; e.g. see [77, 16.9]. Then fg is a nonzero central element of the domain $D[x]$ that is contained in the principal right (left) ideal domain $fD[x]$. Therefore, $D[x]$ is a bounded hereditary Noetherian prime ring.   □

Example 7.57 Let $\mathbb{H}$ be the division ring of Hamiltonian quaternions, and let $\mathbb{R}$ be the field of real numbers. Since the noncommutative division ring $\mathbb{H}$ is finite-dimensional over its center $\mathbb{R}$, it follows from Proposition 7.56 that $\mathbb{H}[x]$ is a bounded principal right (left) ideal domain that is not an automorphism-extendable right or left $D[x]$-module. In particular, $\mathbb{H}[x]$ is a bounded hereditary Noetherian prime ring that is not automorphism-extendable right or left $D[x]$-module.

7.6 General Properties of Endomorphism-Extendable Modules and Rings

Proposition 7.58 *Let R be a ring, Q be a right R-module, and let M be an essential submodule in Q.*

1. *Let X be a module and let $f: X \to Q$ be a homomorphism, and there exists a homomorphism $g: X \to M$ such that f coincides with g on $f^{-1}(M)$. Then $f(X) \subseteq M$.*
2. *If f is an endomorphism of the module Q, and there exists an endomorphism g of the module M such that f coincides with g on $M \cap f^{-1}(M)$, then $f(M) \subseteq M$.*
3. *For any module homomorphism $f: X \to Q$, we have that $f^{-1}(M)$ is an essential submodule in X.*
4. *Q/M is a singular module.*

5. *If M is a nonsingular module, then Q is a nonsingular module and the kernel of each nonzero module homomorphism $f: X \to Q$ is not an essential submodule in X. In particular, Q does not have a nonzero endomorphism with essential kernel.*

Proof 1. We assume that $m = (f - g)(x) \in M \cap (f - g)(M')$, where $x \in X$. Then

$$f(x) = (f - g)(x) + g(x) = m + g(x) \in M, \quad x \in f^{-1}(M),$$

$$m \in (f - g)(f^{-1}(X)) = 0, \quad M \cap (f - g)(X) = 0.$$

Since Q is an essential extension of the module M, $(f - g)(X) = 0$. Therefore, $f(X) = g(X) \subseteq M$.

2. The assertion follows from (1), for $X = M$.

3. Let Y be a nonzero submodule in X. If $f(Y) = 0$, then $0 \neq Y \subseteq f^{-1}(0) \subseteq Y \cap f^{-1}(M)$. We assume that $f(Y) \neq 0$. Then $M \cap f(Y) \neq 0$, whence $0 \neq f^{-1}(M \cap f(Y)) \subseteq f^{-1}(M)$ and $Y \cap f^{-1}(M) \neq 0$.

4. Let $q \in Q$ and let $f: R_R \to Q$ be a homomorphism such that $f(a) = qa$ for all $a \in R$. By (3), $f^{-1}(M)$ is an essential right ideal of the ring R. Since $f^{-1}(M)$ is the annihilator of the element $q + X \in Q/X$, the module Q/M is singular.

5. Since $M \cap Z(Q) = 0$ and Q is an essential extension of M, the module Q is nonsingular. If $f \in \text{Hom}(X, Q)$ and X is an essential extension of $\text{Ker } f$, then it follows from (4) that $Z(N/\text{Ker } f) = N/\text{Ker } f \cong f(N)$, whence $f(N) = Z(f(N)) \subseteq Z(M) = 0$ and f is the zero homomorphism. $\square$

Proposition 7.59 *Let M be a module with injective hull E. The following conditions are equivalent.*

1. *M is a strongly endomorphism-extendable module.*
2. *$f(M) \subseteq M$ for any endomorphism f of the module E that maps into itself from some essential submodule of the module M.*
3. *M is an endomorphism-extendable module, and $h(M) \subseteq M$ for any endomorphism h of the module E such that $\text{Ker } h$ is an essential submodule of E.*

Proof (1) $\Rightarrow$ (2). Let X be an essential submodule of the module M, and let f be an endomorphism of the module Q with $f(X) \subseteq X$. We set $Y = M \cap f^{-1}(M)$. Then $X \subseteq Y$, the endomorphism f induces the homomorphism $g_1: Y \to M$, and $g_1(X) = f(X) \subseteq X$. Since M is a strongly endomorphism-extendable module, the homomorphism g_1 can be extended to an endomorphism g of the module M. By Proposition 7.58(2), $f(M) \subseteq M$.

$(2) \Rightarrow (3)$. We denote by X the essential submodule $M \cap \mathrm{Ker}\, h$ of the module M. Since $h(X) = 0 \subseteq X$, it follows from (2) that $h(M) \subseteq M$. Let X be an essential submodule of M, and let g_1 be an endomorphism of the module X. Since the module E is injective, g_1 can be extended to an endomorphism f of the module E. By (2), f induces the endomorphism g of the module M that is an extension of the endomorphism g_1. By Corollary 7.7, M is an endomorphism-extendable module. $\square$

Proposition 7.60 *Every nonsingular endomorphism-extendable module M is a strongly endomorphism-extendable module.*

Proof By Proposition 7.58(5), the injective hull E of the nonsingular module M does not have nonzero endomorphisms with essential kernels. Therefore, the conclusion follows easily from the above proposition. $\square$

For a module X, an endomorphism f of X is said to be locally nilpotent if $X = \bigcup_{n=1}^{\infty} \mathrm{Ker}(f^n)$; i.e. for any $x \in X$, there exists an integer $n \in \mathbb{N}$ with $f^n(x) = 0$. In this case, for any finitely generated submodule Y in X, there exists an integer $n \in \mathbb{N}$ with $f^n(Y) = 0$.

Proposition 7.61 *Let R be a ring, let M be an endomorphism-extendable right R-module, and let E be the injective hull of the module M.*

1. *If every essential submodule in M is fully invariant in M, then M is a strongly endomorphism-extendable module.*
2. *If for every essential submodule N in M, the factor module M/N is semi-Artinian, then M is a strongly endomorphism-extendable module.*
3. *If for every $m \in M$ and every endomorphism g of the module M with essential kernel, there exists a positive integer n with $g^{n+1}(m) \in \sum_{i=0}^{n} g^i(m)R$, then M is a strongly endomorphism-extendable module.*
4. *If every endomorphism g of the module M with essential kernel is locally nilpotent, then M is a strongly endomorphism-extendable module.*
5. *If M is a locally Noetherian module, then M is a strongly endomorphism-extendable module.*

Proof Let h be an endomorphism of the injective hull E of the module M such that $\mathrm{Ker}\, h$ is an essential submodule. By Proposition 7.58(1), it is sufficient to prove that $h(M) \subseteq M$ in all considered cases 1–5.

Let $P \equiv M \cap h^{-1}(M)$, and let $N \equiv \{m \in M \mid h^n(M) \subseteq M \text{ for all } n \in \mathbb{N}\}$. Since $N \supseteq M \cap \mathrm{Ker}\, h$, we have that E, M are essential extensions of the module N. In addition, N is the largest submodule of M with $h(N) \subseteq N$. Since M is an endomorphism-extendable module and $h(N) \subseteq N$, we have that $(h - g)(N) = 0$ for some $g \in \operatorname{End} M$. Since $g(M \cap \operatorname{Ker} h) = 0$, we have $g \in \Delta(M)$, where $\Delta(M)$ is the set of $f \in \operatorname{End}(M)$ such that $\operatorname{Ker} f$ is essential in M. We set $\overline{M} = M/N$. If $\overline{M} = 0$, then $M = N$, $h(M) \subseteq M$, and the assertion is proved in this case.

Now we assume that $\overline{M} \neq 0$. Then $(h - g)(N) = 0$ and the relation $t(x + N) = (h - g)(x)$ correctly defines a homomorphism $t \colon \overline{M} \to E$. Let $\overline{V} \equiv t^{-1}(N) \subseteq \overline{M}$ and let V be the submodule in M such that $V \supseteq N$ and $V/N = \overline{V}$. Since E is an essential extension of the module M, it follows that $\overline{M}$ is an essential extension of the module $\overline{V}$. Therefore, $\overline{V} \neq 0$. In addition, $h(V) \subseteq t(V) + g(V) \subseteq N + g(V) \subseteq M$. Since $V \supseteq N$, we have that M is an essential extension of the module V.

1. By assumption, $g(V) \subseteq V$. Therefore, $h(V) \subseteq N + g(V) \subseteq N + V = V$. Since N is the largest submodule of M with $h(N) \subseteq N$, we have $V = N$. Therefore, $\overline{V} = 0$. This is a contradiction.

2. Since $h(V) \subseteq M$ and $h(N) \subseteq N$, we have that h induces homomorphism $\overline{h} \colon \overline{V} \to \overline{M}$. Since M is an essential extension of the module V, it follows from the assumption that the nonzero module $\overline{M}$ is an essential extension of its nonzero socle $\overline{S} = S/N$, where $N \in \operatorname{Lat}(S)$. Since $\overline{V}$ is an essential submodule in $\overline{M}$, we have $\overline{S} \subseteq \overline{V}$, whence $\overline{h}(\overline{S}) \subseteq \overline{S}$. Therefore, $h(S) \subseteq S$. Since N is the largest submodule in M with $h(N) \subseteq N$, we have $S = N$. Therefore, $\overline{S} = 0$. This is a contradiction.

3. Let $m \in M \setminus N$, $0 \neq \overline{m} = m + N \in \overline{M}$. By assumption, $g^{n+1}(m) \in \sum_{i=0}^{n} g^i(m)R$ for some n. Since $(h^i - g^i)(N) = 0$ for all i, the relation $t_i(\overline{m}x) = (h^i - g^i)(mx)$, $x \in R$, correctly defines homomorphisms $t_i \colon \overline{m}R \to E$. Let $\overline{V}_i \equiv t_i^{-1}(N) \subseteq \overline{m}R$ and $\overline{W} \equiv \overline{V}_1 \cap \cdots \cap \overline{V}_n \subseteq \overline{m}R$. Since Q is an essential extension of the module M, we have that $\overline{m}R$ is an essential extension of each of the modules $\overline{V}_i$. Therefore, $\overline{m}A$ is an essential extension of the module $\overline{W}$. Then $0 \neq \overline{m}a \in \overline{W}$ for some $a \in A$. Therefore, $h^i(ma) = t_i(ma) + g^i(ma) \in M$ for all $i = 1, \ldots, n$ and $h^{n+1}(ma) \in \sum_{i=0}^{n} h^i(ma)R \in M$. Then $h^{n+j}(ma) \in M$ for all $j \geq 1$. Therefore, $h^k(ma) \in M$ for all $k \geq 1$, $ma \in N$, $\overline{m}a = 0$. This is a contradiction.

4. In this case, we are under conditions of (3).

5. Let X be an arbitrary cyclic submodule of M. By Lemma 7.13, there exists a positive integer $n = n(X, h)$ with $h^n(X) = 0$. Therefore, we are under conditions of (4). $\qquad\qquad\square$

Theorem 7.62 *Let R be a right endomorphism-extendable ring.*

1. *If every essential right ideal of R is an ideal, then R is a right strongly endomorphism-extendable ring.*
2. *If for every essential right ideal N of the ring R, the module R/N is semi-Artinian, then R is a right strongly endomorphism-extendable ring.*
3. *If for every $m \in R$ and each element $g \in Z(R_R)$, there exists a positive integer n with $g^{n+1}(m) \in \sum_{i=0}^{n} g^i(m)R$, then R is a right strongly endomorphism-extendable ring.*
4. *If every element of the ideal $Z(R_R)$ is nilpotent, then R is a right strongly endomorphism-extendable ring.*
5. *If R is a right Noetherian ring, then R is a right strongly endomorphism-extendable ring.*

The preceding theorem follows from Proposition 7.61.

Proposition 7.63 *Let M be a strongly endomorphism-extendable module and $S = \sum_{h \in \Delta(M)} h(M)$.*

1. *The ideal $\Delta(M)$ of the ring End M is contained in the Jacobson radical $J(\text{End } M)$ of the ring End M.*
2. *If M is an essential extension of some quasi-injective module X, then M is a quasi-injective module. In particular, if M is an essential extension of a semisimple module, then M is a quasi-injective module.*
3. *If M is an essential extension of the module S, then M is a quasi-injective module and $M = G(M)$.*
4. *If M is an indecomposable module, then M is a uniform module and either M is a quasi-injective module and $M = G(M)$ or every nonzero endomorphism of the module M is a monomorphism and End M is a domain.*

Proof Let E be the injective hull of the module M.

1. Let $h \in \Delta(M)$. Then $h(X) = 0$ for some essential submodule X in M. Since $\Delta(M)$ is an ideal of the ring End M, it is sufficient to prove that $1 - h$ is an automorphism of the module M. Since the module E is injective, $1_M - h$ can be extended to an endomorphism f of the module E, which acts identically on the essential submodule X of the injective module E. Therefore, $f(E)$ is an injective essential submodule of the module E. Then f is an automorphism of the injective module E. The converse automorphism f^{-1} also acts identically on the essential submodule X of the module E. Therefore, $1_E - f^{-1} \in \Delta(E)$. By Lemma 7.59(1), $(1_E - f^{-1})(M) \subseteq M$. Then $f^{-1}(M) \subseteq M$. Therefore, $M \subset f(M) = (1_M - h)(M) \subseteq M$. Then $1_M - h$ is an automorphism of the module M.

2. It is sufficient to prove that $f(M) \subseteq M$ for every endomorphism f of the module E. Since M is an essential extension of the module X, we have that E is the injective hull of the quasi-injective module X. Since X is a quasi-injective module, $f(X) \subseteq X$. By Lemma 7.59(1), $f(M) \subseteq M$. The second assertion follows from the first assertion and the property that every semisimple module is quasi-injective.

3. By Lemma 7.59(1), $h(M) \subseteq M$ for every $h \in \Delta(E)$. Therefore, $\sum_{h \in \Delta(E)} h(E) = S$ is an essential submodule in M. Then E is the injective hull of the module S. By Theorem 1.16(f), $S \subseteq Z(E)$ and S is a fully invariant invariant submodule of E. Therefore, $M = G(M)$ and S is a quasi-injective module. By (2), M is a quasi-injective module.

4. Since M is an indecomposable quasi-continuous module, M is a uniform module.

If $S \neq 0$, then M is an essential extension of the module S, and it follows from (2) that M is a quasi-injective module and $M = G(M)$.

We assume that $S = 0$; i.e. the kernel of each nonzero endomorphism of the module M is not essential in M. Since M is a uniform module, every nonzero endomorphism of the module M is a monomorphism. Therefore, End M is a domain. $\qquad\square$

Corollary 7.64 *If R is a right strongly endomorphism-extendable ring, then* $Z(R_R) \subseteq J(R)$.

Corollary 7.64 follows from Proposition 7.63(1).

7.7 Annihilators that Are Ideals

The following lemma is straightforward.

Lemma 7.65 *For a ring R and an element $a \in R$, the following are equivalent:*

1. *The right ideal $r(a)$ is an ideal of R.*
2. *$aRb = 0$ for any two elements $a, b \in R$ with $ab = 0$.*
3. *For every element $b \in R$, the left ideal $\ell(b)$ is an ideal in R.*

Definition 7.66 Let n be a positive integer. A module M is said to be n-endomorphism-extendable if every endomorphism of any n-generated submodule in M can be extended to an endomorphism of the module M.

Lemma 7.67 *Let R be an 1-right endomorphism-extendable ring and $a \in R$. The right ideal $r(a)$ is an ideal of the ring R if and only if the left ideal Ra*

*is an ideal. In particular, if a is an arbitrary left nonzero-divisor in R, then
Ra is an ideal.*

Proof We assume that $r(a)$ is an ideal in R and $b \in R$. Then $r(a) \subseteq r(ab)$.
Therefore, $f(a) = ab$ for some epimorphism $f: aR \to abR$. Then $f \in$
$\text{End}(aR)$. By assumption, $ta = f(a) = ab$ for some $t \in R$. Therefore, $Ra \supseteq$
aR and Ra is an ideal.   □

We assume that Ra is an ideal in R and $b \in R$. Then $ab = ca$ for some
$c \in R$. Therefore, $abr_R(a) = car_R(a) = 0$, $br_R(a) \subseteq r_R(a)$ and $r(a)$ is an
ideal.

Lemma 7.68 *A ring R is left duo if and only if R is an 1-right endomorphism-
extendable ring and r(a) is an ideal in R for every element a ∈ R.*

Proof Let R be a left duo ring. It follows from Lemma 7.65 that $r(a)$ is an
ideal in R for every element $a \in R$. Let $f \in \text{End}\, aR$. Then $f(a) = ab$ for
some element $b \in R$ and $f(ax) = f(a)x = abx$ for every $x \in R$. Since
R is a left duo ring, $ab = ca$ for some element $c \in R$. With the use of the
relation $g(y) = cy$, the endomorphism $g \in \text{End}\, R_R$ is defined. In addition,
$g(ax) = cax = abx = f(ax)$ for all $x \in R$. Therefore, g is an extension of
the homomorphism f.
    The converse follows easily from Lemma 7.67.   □

Lemma 7.69 *Let R be a 1-right endomorphism-extendable ring. Then R has
the left classical ring of fractions Q and a $Ra^{-1} \subseteq R$ for any nonzero-divisor
a ∈ R. If R has the classical right ring of fractions, then Q is the two-sided
classical ring of fractions of the ring R. If R is a domain, then R is a left
invariant domain that has the classical left division ring of fractions.*

Proof Let S be the set of all nonzero-divisors in R, $s \in S$, and let $a \in R$.
By Lemma 7.67, $sa = bs$ for some $b \in R$. Therefore, S is a left Ore set.
Therefore, R has the classical left ring of fractions. The inclusion $aRa^{-1} \subseteq R$
follows from Lemma 7.67.
    If R has the classical right ring of fractions, then Q is the two-sided classical
ring of fractions of the ring R, since for any ring that has the classical right ring
of fractions and the classical left ring of fractions, these two rings of fractions
can be naturally identified.
    The last assertion follows easily from the first assertion.   □

Lemma 7.70 *If a 1-right endomorphism-extendable ring R is reduced or right
duo, then R is a left duo ring.*

In the both cases, all right annihilators of elements of R are ideals. Therefore, Lemma 7.70 follows from Lemma 7.68.

Lemma 7.71 *Let R be a finitely right endomorphism-extendable ring. We know that R has the classical left ring of fractions Q and $s R s^{-1} \subseteq R$ for every nonzero-divisor s in R. If $q \in Q$ and $q^{n+1} = q^n a_n + \cdots + q_1 a + a_0$ for some elements $a_0, \ldots, a_n \in R$, then $q \in s^{-1} R s$ for some nonzero-divisor s in R.*

Proof We denote by M the n-generated submodule $\sum_{i=1}^{n} q^i R$ of the module Q_R. Then $q M \subseteq M$. By 1.26, $s M \subseteq R$ for some nonzero-divisor $s \in R$. The right ideal $s M$ of R is finitely generated and $s \in s M$. In addition,

$$q s^{-1} \cdot s M = q M \subseteq M = s^{-1} \cdot s M.$$

Therefore, with the use of the relation $f(x) = s q s^{-1} x$, we can define an endomorphism f of the finitely generated right ideal $s M$ of the ring R. By assumption, f can be extended to an endomorphism g of the module R_R. We set $c = g(1)$. Then $s q = s q s^{-1} \cdot s = f(s) = cs, q = s^{-1} cs \in s^{-1} R s$. $\square$

7.8 Completely Integrally Closed Subrings and Self-Injective Rings

Let Q be a ring, and let R be a unitary subring in Q. The ring R is called a right completely integrally closed subring in Q if R contains every element $q \in Q$ such that $q B \subseteq B$ for some essential right ideal B of the ring R. The ring R is called a right classically completely integrally closed subring in Q if R contains every element $q \in Q$ such that $q^n a \in R$ for some nonzero-divisor a of R and all positive integers n.

Lemma 7.72 *If Q is the maximal right ring of fractions of the ring R and the ring Q is right self-injective, then R is a right completely integrally closed subring in Q if and only if R is a right strongly endomorphism-extendable ring.*

Since the ring Q is right self-injective, it is well known that Q_R is the injective hull of the module R_R and the ring $\operatorname{End} Q_R$ can be naturally identified with the ring Q; e.g. see [96, section 14.4, proposition 4.1]. Therefore, the preceding lemma follows easily from Lemma 7.59(1). As a consequence of Lemma 7.72, we have the following.

Lemma 7.73 *If R is a right nonsingular ring and Q is the maximal right ring of quotients of the ring R, then R is a right completely integrally closed subring in Q if and only if R is a (strongly) right endomorphism-extendable ring.*

Proposition 7.74 *Let R be a semiprime right Goldie ring, and let Q be the semisimple Artinian classical right ring of fractions of the ring R. The following conditions are equivalent.*

1. *R is a classically completely integrally closed right subring in Q.*
2. *R is a completely integrally closed right subring in Q.*
3. *R is a right strongly endomorphism-extendable ring.*
4. *R is an right endomorphism-extendable ring.*

Proof By Theorem 1.27, the semiprime right Goldie ring R is right nonsingular. In addition, the semisimple Artinian ring Q is the maximal right ring of fractions of the ring R by [33, theorem 16.14]. Therefore, the equivalence of (2), (3) and (4) follows from Lemma 7.73.

  (1) $\Rightarrow$ (2). Let B be an an essential right ideal of the ring R, and let q be an element of the ring Q with $qB \subseteq B$. By Theorem 1.27, the essential right ideal B contains some nonzero-divisor b. Since $qB \subseteq B$, we have that $q^n B \subseteq B$ for each nonnegative integer n. Therefore, $q^n b \in B \subseteq R$ for each nonnegative integer n. Since R is a right classically completely integrally closed subring in Q, we have that $q \in Q$ and R is a right completely integrally closed subring in Q.

  (2) $\Rightarrow$ (1). Let q be an element of the ring Q such that $q^n b \in R$ for some nonzero-divisor $b \in R$ and each nonnegative integer n. We denote by B the right ideal $\sum_{n=0}^{\infty} q^n bR$ of the ring R, Since B contains a non-zero-divisor b, we have that B is an essential right ideal by Theorem 1.27. In addition, $qB \subseteq B$. Since R is a right completely integrally closed subring in Q, we have that $q \in Q$ and R is a right classically completely integrally closed subring in Q. $\square$

Theorem 7.75 *For a ring R, the following conditions are equivalent.*

1. *R is a right endomorphism-extendable right nonsingular ring.*
2. *R is a right strongly endomorphism-extendable right nonsingular ring.*
3. *$R = B \times C$, where B is a right self-injective regular ring, C is a left duo Baer reduced ring and C is a completely integrally closed right subring in its maximal right ring of fractions Q.*

Proof The implication (1) $\Rightarrow$ (2) follows from Lemma 7.59(2).

  (2) $\Rightarrow$ (3). Because of Theorem 1.63, every endomorphism-extendable module is quasi-continuous. Therefore, it follows from Theorem 1.68 that $R = B \times C$, where B is a right self-injective regular ring and C is a right strongly

endomorphism-extendable Baer reduced ring. By Lemma 7.73, C is a right completely integrally closed subring in Q. By Lemma 7.68, C is a left duo ring.

The implication $(3) \Rightarrow (1)$ follows from Lemma 7.73. □

Corollary 7.76 *For a ring R, the following conditions are equivalent.*

1. *R is a right endomorphism-extendable, right nonsingular indecomposable ring.*
2. *R is a right strongly endomorphism-extendable, right nonsingular indecomposable ring.*
3. *Either R is a right self-injective regular indecomposable ring or R is a left duo, right and left Ore domain that is a right classically completely integrally closed subring in its classical division ring of fractions Q.*

Proof The implication $(3) \Rightarrow (1)$ follows from Proposition 7.74.

The implication $(1) \Rightarrow (2)$ follows from Lemma 7.59(1).

$(2) \Rightarrow (3)$. By Theorem 7.75, R is either a right self-injective regular indecomposable ring or a left duo indecomposable Baer reduced ring that is a right completely integrally closed subring in its maximal right ring of fractions Q. It is sufficient to consider only the second case. The left duo regular indecomposable Baer ring R is a left duo domain. By Theorem 1.68, R is a right and left Ore domain. Let Q be the classical division ring of fractions of the domain R. By Proposition 7.74, R is a right classically completely integrally closed subring in Q. □

Theorem 7.77 *For a ring R, the following conditions are equivalent.*

1. *R is a right strongly endomorphism-extendable ring without nontrivial idempotents.*
2. *Either R is a right self-injective right uniform local ring and $J(R) = Z(R_R)$ is a nonzero ideal or R is a left duo, right and left Ore domain that is a right classically completely integrally closed subring in its classical division ring of fractions.*

Proof $(1) \Rightarrow (2)$. Since R is a ring without nontrivial idempotents, R_R is an indecomposable module. By Proposition 7.63(4), R is a right uniform ring and either R is right self-injective and $R_R = G(R_R)$ or R is a domain. By Corollary 7.76, it is sufficient to consider the case, where the ring R is right self-injective. Then $J(R) = Z(R_R)$ [34, theorem 19.27]. Since R is a right self-injective ring without nontrivial idempotents, R is a local ring.

The implication $(2) \Rightarrow (1)$ follows from Corollary 7.76. □

Theorem 7.78 *If R is a right endomorphism-extendable regular ring, then R is a right self-injective ring.*

Proof Since R is a right endomorphism-extendable ring, R is a right quasi-continuous ring by Theorem 1.63. By Theorem 1.68, R is a direct product of the right self-injective regular ring and a regular ring. Therefore, without loss of generality, we can consider only the case, where every principal right ideal of the ring R is generated by a central idempotent. It is sufficient to prove that for an arbitrary right ideal B of the ring R, every homomorphism $f: B_R \to R_R$ can be extended to a homomorphism $R_R \to R_R$. Let $b \in B$. Then $bR = eR$ for some central idempotent $e \in bR \subseteq B$, Then $f(b) = f(be) = f(b)e = ef(b) \in B$. Therefore, $f(B) \subseteq B$. Since R_R is an endomorphism-extendable module, the homomorphism $f: B_R \to R_R$ can be extended to a homomorphism $R_R \to R_R$.   □

Proposition 7.79 *Let R be a right strongly endomorphism-extendable ring. If R has an essential quasi-injective right ideal, then the ring R is right self-injective. In particular, if R has an essential semisimple right ideal, then the ring R is right self-injective. If $Z(R_R)$ is an essential right ideal of the ring R, then the ring R is right self-injective.*

The preceding proposition follows from Propositions 7.63(2), 7.63(3) and 7.74(3).

Theorem 7.80 *Every endomorphism-extendable semi-Artinian module M is quasi-injective. In particular, every right endomorphism-extendable right semi-Artinian ring is right self-injective.*

Proof Let Q be the injective hull of the module M, S be the socle of the module Q, and let f be an endomorphism of the module Q. It is sufficient to prove that $f(M) \subseteq M$. We remark that $f(S) \subseteq S$. In addition, $S \subseteq M$, since M is an essential submodule in Q.

We denote by X the sum of all submodules X' in M with $f(X') \subseteq X'$. Then $f(X) \subseteq X$, $S \subseteq X$, and X is an essential submodule in M.

If $X = M$, then $f(M) \subseteq M$, which is required.

We assume that $X \neq M$. Then the nonzero semi-Artinian module M/X is an essential extension of its nonzero socle Y/X, where $X \subsetneq Y \subseteq M$. Since $f(X) \subseteq X$ and M is an endomorphism-extendable module, there exists an endomorphism g of the module M with $(f - g)(X) = 0$. In addition, Y/X is a semisimple module. Therefore, $(f - g)(Y) \subseteq S \subseteq X$. Since $f(X) \subseteq X)$, f induces the endomorphism $\overline{f}$ of the module M/X with nonzero socle Y/X. Since Y/X is the socle of the module M/X, we have that $\overline{f}(Y/X) \subseteq Y/X$

and $f(X) \subseteq X$. Therefore, $f(Y) \subseteq Y$. In addition, Y properly contains X; this contradicts to the choice of X. □

7.9 Endomorphism-Liftable and π-Projective Modules

Definition 7.81 A module M is said to be endomorphism-liftable if for every epimorphism $h: M \to \overline{M}$ and each endomorphism $\bar{f}$ of the module $\overline{M}$, there exists an endomorphism f of the module M with $\bar{f}h = hf$.

Definition 7.82 A module M is said to be π-projective if for every epimorphism $h: M \to \overline{M}$ and each idempotent endomorphism $\bar{f}$ of the module $\overline{M}$, there exists an endomorphism f of the module M with $\bar{f}h = hf$.

1. Every quasi-projective module is endomorphism-liftable. Every quasi-cyclic abelian group $\mathbb{Z}(p^\infty)$ is an endomorphism-liftable non-quasi-projective $\mathbb{Z}$-module.
2. Every uniserial module is π-projective. The ring $\mathbb{Z}$ is a π-projective non-uniserial $\mathbb{Z}$-module.
3. Every endomorphism-liftable module is π-projective. If R is a noncomplete discrete valuation domain and Q is the field of fractions of the domain R, then Q is a π-projective R-module that is not endomorphism-liftable.

The preceding assertions are not too difficult to verify.

Lemma 7.83 *Let M be a* π*-projective module and let X, Y be two submodules in M with* $X + Y = M$. *Then there exist homomorphisms* $f: M \to X$ *and* $g: M \to Y$ *such that*

$$f(Y) + g(X) \subseteq X \cap Y, \quad (f + g - 1_M)(M) \subseteq X \cap Y,$$
$$M = (f + g)(M) + X \cap Y, \quad \mathrm{Ker}(f + g) \subseteq X \cap Y.$$

Proof Let $h: M \to M/(X \cap Y)$ be the natural epimorphism. Since $h(M) = h(X) \oplus h(Y)$, there exist natural projections $\bar{f}: h(M) \to h(X)$ and $\bar{g}: h(M) \to h(Y)$. Since M is π-projective, $\bar{f}h = hf$ and $\bar{g}h = hg$ for some $f, g \in \mathrm{End}\, M$. Therefore,

$$f(M) \subseteq X, \quad g(M) \subseteq Y, \quad f(Y) + g(X) \subseteq X \cap Y.$$

Since $(\bar{f} + \bar{g} - 1_{h(M)}) = 0$, we have $(f + g - 1_M)(M) \subseteq X \cap Y$. Since $M = (f + g)(M) + (f + g - 1_M)(M)$, we have $M = (f + g)(M) + X \cap Y$. If $x \in \mathrm{Ker}(f + g)$, then $x = (1_M - f - g)(x) \in X \cap Y$. Then $\mathrm{Ker}(f + g) \subseteq X \cap Y$. □

Lemma 7.84 *For a module M, the following conditions are equivalent.*

1. *M is an endomorphism-liftable module, and every factor module of M is endomorphism-extendable.*
2. *M is an endomorphism-extendable module, and every submodule of M is endomorphism-liftable.*

Proof $(1) \Rightarrow (2)$. Let N be a submodule in M, $\overline{f}$ be an endomorphism of the factor module $\overline{N} = N/P$, and let $h: M \to M/P$ be the natural epimorphism. Since M/P is an endomorphism-extendable module, $\overline{f}$ can be extended to an endomorphism $\overline{g}$ of the module M/P. Since M is an endomorphism-liftable module, $\overline{g}h = hg$ for some $g \in \operatorname{End} M$. Therefore, $g(N) \subseteq N$. Then g induces $f \in \operatorname{End} N$ and $\overline{f}h_N = h_N f$, where $h_N: N \to N/P$ is the natural epimorphism. Therefore, N is an endomorphism-liftable module.

$(2) \Rightarrow (1)$. Let M/P an arbitrary factor module, $h: M \to M/P$ be the natural epimorphism, let N/P be a submodule in M/P, let h_N be the restriction of h to N, and let $\overline{f}$ be an endomorphism of the module N/P. By assumption, N is an endomorphism-liftable module. Therefore, there exists an endomorphism f of the module N with $\overline{f}h_N = h_N f$. Since M is an endomorphism-extendable module, the endomorphism f can be extended to an endomorphism g of the module M/P; in addition, $g(P) = f(P) \subseteq P$. Since $g(P) \subseteq P$, we have that g induces the endomorphism $\overline{g}$ of the module M/P that is an extension of the endomorphism $\overline{f}$ of the module N/P. Therefore, M/P is an endomorphism-extendable module. $\qquad\square$

7.10 Rings Whose Cyclic Modules Are Endomorphism-Extendable

It is known that if R is a ring over which all cyclic right modules are quasi-continuous, then $R = B \times C$, where B is a semisimple Artinian ring, $C = C_1 \times \cdots \times C_n$ and all the rings C_i are right uniform. The ring C_i is local if and only if C_i is a right uniserial ring. If all cyclic right R-modules are quasi-injective, then for every i, we have that C_i is a right self-injective, right and left uniserial, right and left invariant ring and $J(C_i)$ is a nil-ideal (see e.g. [67]).

Proposition 7.85 *For a ring R, the following conditions are equivalent.*

1. *Every cyclic right R-module is endomorphism-extendable.*
2. *R is a right endomorphism-extendable ring and every right ideal of the ring R is endomorphism-liftable.*

3. $R = B \times C$, where B is a semisimple Artinian ring, $C = C_1 \times \cdots \times C_n$ and all the rings C_i are right uniform rings, in which all right ideals are endomorphism-liftable.
4. $R = B \times C$, where B is a semisimple Artinian ring, $C = C_1 \times \cdots \times C_n$ and all the rings C_i are right uniform rings, over which all cyclic modules are endomorphism-extendable.

Proof The projective module R_R is endomorphism-liftable. Therefore, the equivalences (1) $\Leftrightarrow$ (2) and (3) $\Leftrightarrow$ (4) follow from Lemma 7.84.

The implication (4) $\Rightarrow$ (1) is not difficult to check.

(1) $\Rightarrow$ (4). Since every endomorphism-extendable module is quasi-continuous, all cyclic right R-modules are quasi-continuous. By [67], $R = B \times C$, where B is a semisimple Artinian ring, $C = C_1 \times \cdots \times C_n$, and all the rings C_i are right uniform. It is clear that all cyclic right C_i-modules are endomorphism-extendable. $\qquad\square$

Proposition 7.86 *For a ring R, the following conditions are equivalent.*

1. *R is a right strongly endomorphism-extendable ring, and every cyclic right R-module is endomorphism-extendable.*
2. *R is a right strongly endomorphism-extendable ring, and every right ideal of the ring R is endomorphism-liftable.*
3. *$R = A_1 \times \cdots \times A_n$ and for every A_i, each right ideal of the ring A_i is endomorphism-liftable and either A_i is a simple Artinian ring or A_i is a right uniserial right self-injective ring and $J(R)$ coincides with nonzero ideal $Z(R_R)$, or A_i is a left invariant, right and left Ore domain that is a right classically completely integrally closed subring in its classical division ring of fractions.*

Proof The equivalence of (1) and (2) follows from Proposition 7.85.

The implication (3) $\Rightarrow$ (2) follows from Corollary 7.76.

(2) $\Rightarrow$ (3). By Proposition 7.85, it is sufficient to consider the case, where R is a right uniform ring.

If R is a domain, then it follows from Corollary 7.76 that R is a left duo, right and left Ore domain that is a right classically completely integrally closed subring in its classical division ring of fractions.

We assume that R is not a domain. Since R is a right uniform ring with left zero-divisors, $Z(R_R)$ is an essential right ideal of the ring R. By Theorem 7.77, R is a right self-injective, right uniform local ring and $J(R) = Z(R_R)$ is a nonzero ideal. Clearly, then R is a right uniserial ring. $\qquad\square$

A ring R is said to be integral over its center if for every element $s \in R$, there exist central elements $c_1, \ldots, c_n$ of the ring R with $s^{n+1} = \sum_{i=0}^{n} s^i c_i$.

Corollary 7.87 *If* R *is a ring integral over its center, then the following conditions are equivalent.*

1. R *is a right strongly endomorphism-extendable ring, and every cyclic right R-module is endomorphism-extendable.*
2. R *is a right strongly endomorphism-extendable ring, and every right ideal of the ring R is an endomorphism-liftable.*
3. $R = A_1 \times \cdots \times A_n$ *and for every A_i, each right ideal of the ring A_i is endomorphism-liftable, and either A_i is a simple Artinian ring or A_i is a right uniserial right self-injective ring and $J(R)$ coincides with nonzero ideal $Z(R_R)$, or A_i is a left duo, right and left Ore domain that is a right classically completely integrally closed subring in its classical division ring of fractions.*

Corollary 7.87 follows from Proposition 7.86 and Theorem 7.62(3).

Theorem 7.88 *Let R be a ring such that all essential right ideals of R are ideals. The following conditions are equivalent.*

1. *Every cyclic right R-module is endomorphism-extendable.*
2. $R = A_1 \times \cdots \times A_n$ *and for every A_i, each right ideal of the ring A_i is endomorphism-liftable, and either A_i is a simple Artinian ring or A_i is an right duo right uniserial right self-injective ring and $J(R)$ coincides with the nonzero ideal $Z(R_R)$, or A_i is a duo domain that is a right classically completely integrally closed subring in its classical division ring of fractions.*

Proof The implication (2) $\Rightarrow$ (1) follows from Proposition 7.86.

(1) $\Rightarrow$ (2). By Proposition 7.85, it is sufficient to consider the case, where R is a right uniform ring. Then all nonzero right ideals of the ring R are essential. Since all essential right ideals of the ring R are ideals, the right uniform ring R is a right duo ring. By Theorem 7.62(1), R is a right strongly endomorphism-extendable ring. Now we use Proposition 7.86. $\qquad\square$

Lemma 7.89 *Let R be a unitary subring of the ring B, M be a submodule of the module B_R, and let there exist elements $m_1, \ldots, m_n \in M$, $b_1, \ldots, b_n \in B$ such that $1 = \sum_{i=1}^{n} m_i b_i$ and $b_i M \subseteq R$ for all i. Then $M = \sum_{i=1}^{n} m_i R$ is a projective n-generated module.*

Proof Let $f_1, \ldots, f_n \colon M_A \to R_R$ be homomorphisms such that $f(m) = b_i m$ for $m \in M$. Then $m = \sum_{i=1}^{n} m_i f_i(m)$ for every $m \in M$, and the assertion follows from Theorem 1.58. $\qquad\square$

Proposition 7.90 *Let R be a right endomorphism-extendable ring such that all right ideals of R are π-projective.*

1. *For any two right ideals B and C of R, there exist elements $x, y \in R$ such that*

$$x(B+C) \subseteq B, \quad y(B+C) \subseteq C, \quad xB + yC \subseteq B \cap C,$$

$$(x+y-1)(B+C) \subseteq B \cap C, \quad B+C = (x+y)(B+C) + B \cap C,$$

$$(B+C) \cap r(x+y) \subseteq B \cap C.$$

2. *For any two right ideals B and C of the ring R, there exist elements $s, t \in R$ such that $s + t = 1$ and $sB + tC \subseteq B \cap C$.*
3. *If $d_1, \dots, d_n$ are nonzero-divisors of the ring R and $D = \sum_{i=1}^{n} Rd_i$, then finitely generated left ideal D is a projective left R-module.*
4. *If R is a domain, then R is a left invariant, right semihereditary and left, right and left Ore domain.*

Proof 1. We set $M = B + C$. Since M_R is a π-projective module, it follows from Lemma 7.83 that there exist homomorphisms $f: M \to B$ and $g: M \to C$ such that

$$f(B) + g(C) \subseteq B \cap C, \quad (f+g-1_M)(M) \subseteq B \cap C,$$

$$M = (f+g)(M) + B \cap C, \quad \mathrm{Ker}(f+g) \subseteq B \cap C.$$

Since f, g are endomorphisms of the right ideal M of the right endomorphism-extendable ring R, there exist elements $x, y \in R$ such that $f(m) = xm \in B$ and $g(m) = ym \in C$ for all $m \in M$. Therefore,

$$xM \subseteq B, \quad yM \subseteq C, \quad xB + yC \subseteq B \cap C,$$

$$(x+y-1)M \subseteq B \cap C, \quad M = (x+y)M + B \cap C,$$

$$M \cap r(x+y) \subseteq B \cap C.$$

2. By (1), there exist elements $x, y \in R$ such that $xB + yC \subseteq B \cap C$ and $(x+y-1)(B+C) \subseteq B \cap C$. We set $s = y$ and $t = 1 - y = x - (x+y-1)$. Then

$$s + t = 1, \quad sB + tC \subseteq yB + xC + (x+y-1)C \subseteq B \cap C.$$

3. By Lemma 7.69, R has the classical left ring of fractions Q and $d_i R d_i^{-1} \subseteq R$ for every non-zero-divisor d_i. Therefore, $d_1 \cdots d_n d_i^{-1} \equiv a_i \in R$ for all i. Each the element a_i is a nonzero-divisor in R. Therefore,

$a_i^{-1} \in Q$. By (a), for right ideals $a_{n-1}R$ and $a_n R$, there exists an element $t_{n-1} \in R$ such that

$$(1 - t_{n-1})a_{n-1}R + t_{n-1}a_n R \equiv B_{n-1} \subseteq a_{n-1}R \cap a_n R.$$

By (1), for the right ideals $a_{n-2}R$ and B_{n-1}, there exists an element $t_{n-2} \in R$ such that

$$(1 - t_{n-2})a_{n-2}R + t_{n-2}B_{n-1} \equiv B_{n-2} \subseteq a_{n-2}R \cap B_{n-1}$$

$$\subseteq a_{n-2}R \cap a_{n-1}R \cap a_n R.$$

We assume that $k < n - 1$ and we have a right ideal $B_{n-k} \subseteq \cap_{i=0}^{k} a_{n-i} R$. It follows from (1) that for the right ideals $a_{n-k-1}R$ and B_{n-k}, there exists an element $t_{n-k-1} \in R$ such that

$$(1 - t_{n-k-1})a_{n-k-1}R + t_{n-k-1}B_{n-k}$$

$$\equiv B_{n-k-1} \subseteq a_{n-k-1}R \cap B_{n-k} \subseteq \cap_{i=0}^{k} a_{n-i} R.$$

Finally, there exists an element $t_1 \in R$ such that

$$B_1 = (1 - t_1)a_1 R + t_1 B_2 \subseteq a_1 R \cap B_2 \subseteq \cap_{i=0}^{n-1} a_{n-i} R = \cap_{i=1}^{n} a_i R,$$

$$B_1 = (1 - t_1)a_1 R + t_1((1 - t_2)a_2 R + t_2((1 - t_3)a_3 R + \ldots t_{n-1}a_n R)\ldots)$$

$$= (1 - t_1)a_1 R + t_1(1 - t_2)a_2 R + t_1 t_2(1 - t_3)a_3 R + \cdots$$

$$+ t_1 t_2 \ldots t_{n-2}(1 - t_{n-1})a_n R + t_1 t_2 \ldots t_{n-1}a_n R \subseteq \cap_{i=1}^{n} a_i R,$$

$$(1 - t_1)a_1 \in \cap_{i=1}^{n} a_i R, \quad t_1(1 - t_2)a_2 \in \cap_{i=1}^{n} a_i R,$$

$$t_1 t_2 (1 - t_3)a_3 \in \cap_{i=1}^{n} a_i R, \quad \ldots,$$

$$t_1 t_2 \ldots t_{n-1}a_n \in \cap_{i=1}^{n} a_i R.$$

Let

$$(1 - t_1)a_1 = a_i f_{1i} = b_1, \quad t_1(1 - t_2)a_2 = a_i f_{2i} = b_2, \quad \ldots,$$

$$t_1 t_2 \ldots t_{n-1}a_n = a_i f_{ni} = b_n, \quad M \equiv \sum_{i=1}^{n} Ra_i^{-1} \subseteq Q.$$

Then $\sum_{i=1}^{n} b_i a_i^{-1} = 1$. In addition, for every b_j, we have

$$Mb_j = \sum_{i=1}^{n} Ra_i^{-1}b_j = \sum_{i=1}^{n} Ra_i^{-1}a_i f_{ji} = \sum_{i=1}^{n} Rf_{ji} \subseteq R \subseteq M.$$

By Lemma 7.89, the module $_RM$ is projective. Therefore,

$$_RD = \sum_{i=1}^{n} Rd_i(d_1 \cdots d_n)^{-1}d_1 \cdots d_n = \sum_{i=1}^{n} Ra_i^{-1}d_1 \cdots d_n$$

$$= {}_RMd_1 \cdots d_n \cong {}_RM,$$

whence $_RD$ is a projective module.

4. By (3), R is a left semihereditary domain. By Corollary 7.76, R is a left duo, right and left Ore domain with classical division ring of fractions Q. By Theorem 1.93(3), T is a right semihereditary domain. □

Theorem 7.91 *For a ring R, the following conditions are equivalent.*

1. *R is a right Noetherian ring, and all cyclic right R-modules are endomorphism-extendable.*
2. *R is a left Noetherian ring, and all cyclic left R-modules are endomorphism-extendable.*
3. *$R = A_1 \times \cdots \times A_n$, where A_i is a simple Artinian ring or a uniserial Artinian ring or a duo hereditary Noetherian domain, $i = 1, \ldots, n$.*

Proof It is sufficient to prove the equivalence of conditions (1) and (3).

$(1) \Rightarrow (3)$. By Theorem 7.62(5), R is a right strongly endomorphism-extendable ring. By Proposition 7.86, it is sufficient to consider the case, where either R is a right uniserial right self-injective ring, or R is a left duo domain.

We assume that R is a right uniserial right self-injective ring. Since R is a right Noetherian right self-injective ring, R is an Artinian ring [34, theorem 24.5]. By Theorem 7.80, all cyclic right R-modules are quasi-injective. By [24, theorem 14.7], R is a uniserial duo ring.

We assume that R is a left duo domain. Since R is a right Noetherian, left duo ring, R is a left Noetherian ring. By Proposition 7.90(4), R is a right and left semihereditary domain. The right and left Noetherian, right and left semihereditary domain R is a hereditary domain. Thus, R is a duo hereditary Noetherian domain.

$(3) \Rightarrow (1)$. By Proposition 7.85, we can assume that R is either a uniserial Artinian ring or an invariant hereditary Noetherian domain. Let M be a nonzero cyclic right R-module. If R is a uniserial Artinian ring, then M is an endomorphism-extendable module by Theorem 7.21.

We assume that R is an invariant hereditary Noetherian domain. If M is a nonsingular module, then M is a uniform module and M is an endomorphism-extendable module by Theorem 7.54. If the cyclic module M is not nonsingular, then M has the nonzero annihilator B. By Lemma 7.53(1), R/B is a finite direct product of invariant uniserial Artinian rings. By Theorem 7.21, M is a quasi-injective R/B-module. Therefore, M is a quasi-injective R-module. In particular, M is an endomorphism-extendable module.   □

Notes

The results of this chapter are taken from [121], [123] and [128].

8

Automorphism-Liftable Modules

In this chapter, we study automorphism-liftable modules. Among other things, we describe all automorphism-liftable torsion modules over non-primitive hereditary Noetherian prime rings. We also study automorphism-liftable non-torsion modules over not necessarily commutative Dedekind prime rings.

8.1 Automorphism-Liftable and Endomorphism-Liftable Modules

Definition 8.1 A module M is said to be automorphism-liftable (resp., strongly automorphism-liftable) if, for any epimorphism $h\colon M \to \overline{M}$ and every automorphism $\bar{f}$ of the module $\overline{M}$, there exists an endomorphism (resp., automorphism) f of the module M with $\bar{f}h = hf$.

The notion of an automorphism-liftable module is dual to the notion of an automorphism-extendable module studied in [116] and [128]. We recall that a module M is said to be automorphism-extendable if every automorphism of any its submodule can be extended to an endomorphism of the module M.

Definition 8.2 A module M is said to be endomorphism-liftable if, for any epimorphism $h\colon M \to \overline{M}$ and every endomorphism $\bar{f}$ of the module $\overline{M}$, there exists an endomorphism f of the module M with $\bar{f}h = hf$.

In the preceding definitions, without loss of generality, we can assume that $\overline{M}$ is an arbitrary factor module of the module M and $h\colon M \to \overline{M}$ is the natural epimorphism.

It is clear that all strongly automorphism-liftable modules and all endomorphism-liftable modules are automorphism-liftable.

Endomorphism-liftable modules and abelian groups were studied in many papers under various names; e.g. see [68], [82], [102], [104], [105], [128]. In particular, endomorphism-liftable abelian groups were studied in [82] and [68]; endomorphism-liftable modules over non-primitive hereditary Noetherian prime rings were studied in [104], [105].

Any quasi-cyclic abelian group $\mathbb{Z}(p^{\infty})$ is an endomorphism-liftable (automorphism-liftable) non-quasi-projective $\mathbb{Z}$-module. The ring of integers $\mathbb{Z}$ is an automorphism-liftable $\mathbb{Z}$-module that is not strongly automorphism-liftable. Indeed, let $\bar{f}$ be an automorphism of the simple $\mathbb{Z}$-module $\mathbb{Z}/5\mathbb{Z}$ such that $\bar{f}$ multiplies all elements of this module by 3. Since the only non-identity automorphism of the module $\mathbb{Z}_{\mathbb{Z}}$ coincides with the multiplication by -1, the projective module $\mathbb{Z}_{\mathbb{Z}}$ is not strongly automorphism-liftable.

Mishina [82] completely described strongly automorphism-liftable abelian groups, i.e. strongly automorphism-liftable $\mathbb{Z}$-modules. It follows from this description that strongly automorphism-liftable $\mathbb{Z}$-modules are torsion automorphism-liftable $\mathbb{Z}$-modules. In [4], torsion automorphism-liftable modules[1] are described, and it follows from this description and the results of Mishina [82] that the strongly automorphism-liftable $\mathbb{Z}$-modules coincide with torsion automorphism-liftable $\mathbb{Z}$-modules.

8.2 Non-Primitive Hereditary Noetherian Prime Rings

A ring R is said to be **right bounded** (resp., **left bounded**) if every its essential right (resp., left) ideal contains a nonideal ideal of the ring R. If R is a non-primitive hereditary Noetherian prime ring, then R is not a right or left Artinian; see [80]. Every hereditary Noetherian prime ring R is a primitive ring or a bounded ring, and if R is a primitive bounded ring, then R is a simple Artinian ring; see [80].

Let R be a Noetherian prime ring. It is well known that the ring R has the simple Artinian classical ring of fractions Q. An ideal I of the ring R is called an **invertible** ideal if there exists a sub-bimodule I^{-1} of the bimodule $_RQ_R$ such that $II^{-1} = I^{-1}I = R$.

The maximal elements of the set of all proper invertible ideals of the ring R are called **maximal invertible ideals** of the ring R. The set of all maximal invertible ideals of the ring R is denoted by $\mathbb{P}(R)$. If $P \in \mathbb{P}(R)$, then the submodule $\{m \in M \mid mP^n = 0, n = 1, 2, \ldots\}$ is called the P-primary

---

[1] In that paper, automorphism-liftable modules are called dually automorphism-extendable.

component of the module M; it is denoted by $M(P)$. If $M = M(P)$ for some $P \in \mathbb{P}(R)$, then M is called a primary module or a P-primary module.

We would like to first generalize the description of torsion automorphism-liftable $\mathbb{Z}$-modules from [4] for the case of singular modules over non-primitive hereditary Noetherian prime rings.[2] Note that when we say that R is a non-primitive ring, we mean that R is not a right and left primitive ring, and similarly, when we say that R is Noetherian, then we mean both R_R and $_R R$ are Noetherian.

In order to state and prove our main results of this chapter, we begin with some preparatory lemmas.

Lemma 8.3 *If R is a ring and M is a right R-module, then M is an automorphism-liftable (resp., idempotent-lifted, quasi-projective) R-module if and only if M is an automorphism-liftable (resp., idempotent-lifted; quasi-projective) $R/r(M)$-module.*

The proof of the above lemma is straightforward.

Lemma 8.4 *Let $M = \oplus_{i \in I} M_i$ be a module and $N = \oplus_{i \in I}(N \cap M_i)$ for any submodule N of the module M.*

1. *$\mathrm{Hom}(M_i, M_j) = 0$ for any $i \neq j$ in I; therefore, all the M_i are fully invariant submodules in M.*
2. *If N, P and Q are three submodules of the module M, then the relation $N = P + Q$ is equivalent to the property that $N \cap M_i = P \cap M_i + Q \cap M_i$ for all $i \in I$.*
3. *The module M is automorphism-liftable if and only if each of the modules M_i are automorphism-liftable.*

Proof The assertions (1) and (2) are proved in [114, lemma 2.1(1), (2)]. The assertion (3) follows from (1) and (2). $\qquad\qquad\qquad\qquad\Box$

Lemma 8.5 *[4, Proposition 6, lemma 2] If an automorphism-liftable module M is the direct sum of two modules X and Y, then X, Y are automorphism-liftable modules that are projective with respect to each other.*

Proof It is directly verified that any direct summand of an automorphism-liftable module is automorphism-liftable.

Let A be a submodule of Y and $f : X \to Y/A$ any homomorphism. We prove that f can be lifted to Y. We denote by $\pi : M \to M/A$ the natural projection, we set $\overline{M} = M/A$ and $\overline{S} = (S + A)/A$ for any $S \leq Y$. Then $\overline{M} = \overline{X} \oplus \overline{Y}$. In

---

[2] Over any Noetherian prime ring R, the singular modules coincide with the torsion modules.

our case, we have a natural isomorphism $\overline{X} = (X + A)/A \cong X$, since $A \subset Y$; consequently, $X \cap A = 0$. We have a homomorphism $\overline{f} \colon \overline{X} \to \overline{Y}$ that acts by the rule $x + A \to f(x) + A$.

Now $\overline{M} = (\iota + \overline{f})(\overline{X}) \oplus \overline{Y}$, where $\iota \colon \overline{X} \to \overline{M}$ is the inclusion map. Then the map $g \colon \overline{M} \to \overline{M}$ with $g = (\iota + \overline{f}) \oplus 1_{\overline{Y}}$ is an automorphism of $\overline{M}$, which lifts to an endomorphism h of M, so that, in particular, for every $x \in X$, we have $h(x) + A = x + A + f(x)$, whence $h(x) - x + A = f(x) \in \overline{Y}$, and thus, $h(x) - x \in A \subseteq Y$. So if $h(x) = x' + y'$ $(x' \in X', y' \in Y)$, we must have $x = x'$, and hence

$$f(x) = h(x) - x + A = x + y' - x + A = y' + A = (\pi \circ \eta_Y \circ h)(x),$$

where $\eta_Y \colon X \oplus Y \to Y$ is the canonical projection. Therefore, $f = \pi \circ (\eta_Y \circ h \mid_X)$. It shows that X is Y-projective. The X-projectivity of Y is proved similarly.   □

Lemma 8.6 *If R is a non-primitive hereditary Noetherian prime ring and M is a primary right R-module, then the following conditions are equivalent.*

1. *M is an automorphism-liftable module.*
2. *M is an endomorphism-liftable module.*
3. *For any direct decomposition $M = M_1 \oplus M_2$, the module M_1 is projective with respect to the module M_2.*
4. *M is a projective $R/r(M)$-module or an indecomposable injective R-module.*

Proof The equivalence of conditions (2), (3) and (4) is proved in [105, lemma 13].

(1) $\Rightarrow$ (2) follows from Lemma 8.5, and (2) $\Rightarrow$ (1) is straightforward.   □

Theorem 8.7 *Let R be a non-primitive hereditary Noetherian prime ring, let M be a torsion right R-module, and let $\{M_i\}$ be the set of all primary components of the module M. Then the module M is automorphism-liftable if and only if every primary component M_i of the module M is a projective $R/r(M)$-module or an indecomposable injective R-module.*

Proof For an arbitrary torsion R-module M and every its submodule N, we have $N = \oplus_{i \in I}(N \cap M_i)$; see [114, lemma 2.2(1)]. Therefore, our assertion follows from Lemma 8.6.   □

Remark 8.8 *If R is a non-primitive hereditary Noetherian prime ring, then the structure of indecomposable injective torsion R-modules is known; e.g.*

see [90] and [91]. Namely, the indecomposable injective torsion R-modules coincide with the primary modules M such that all proper submodules of M are cyclic uniserial primary modules and form a countable chain

$$0 = x_0 R \subsetneqq x_1 R \subsetneqq \cdots,$$

all subsequent factors of this chain are simple modules and there exists a positive integer n such that $x_k R \cong x_{k+n} R / x^n R$ and $M(P)/x_k R \cong M(P)/x_{k+n} R$ for all $k = 0, 1, 2, \ldots$.
  If R is a non-primitive Dedekind prime ring, then $n = 1$.

Remark 8.9 *If R is a non-primitive Dedekind prime ring and M is a torsion R-module, then M is quasi-projective if and only if every primary component of the module M is a direct sum of isomorphic cyclic modules of finite length [91, theorem 15].*

Now we are ready to state and prove the first main theorem of this chapter.

Theorem 8.10 *[127] If R is a non-primitive hereditary Noetherian prime ring and M is a singular right R-module, then M is automorphism-liftable if and only if every P-primary component $M(P)$ of the module M is either a projective $R/r(M(P))$-module or a uniserial injective module $M(P)$ such that all proper submodules are cyclic and form a countable chain*

$$0 = x_0 R \subsetneqq x_1 R \subsetneqq \cdots,$$

all subsequent factors of this chain are simple modules and there exists a positive integer n such that $x_k R \cong x_{k+n} R / x_n R$ and $M(P)/x_k R \cong M(P)/x_{k+n} R$ for all $k = 0, 1, 2, \ldots$.

Proof The proof follows from Theorem 8.7, and Remark 8.8. □

In [105], arbitrary endomorphism-liftable modules over non-primitive hereditary Noetherian prime rings are described.

8.3 Non-Primitive Dedekind Prime Rings

A hereditary Noetherian prime ring R is called a Dedekind prime ring (see [81, §5.2]) if any nonzero ideal of R is invertible in the simple Artinian ring of fractions of the ring R. Any hereditary Noetherian prime PI ring is a bounded ring. In particular, all commutative Dedekind domains (e.g. the ring $\mathbb{Z}$)

and full matrix rings over commutative Dedekind domains are non-primitive Dedekind prime rings. Other examples of Dedekind prime rings are given in [81, § 5.2, § 5.3].

The second main result of this section is the next theorem, where the description of singular automorphism-liftable modules is specified in the case where R is a non-primitive Dedekind prime ring.

Theorem 8.11 *[127] If R is a non-primitive Dedekind prime ring and M is a singular right R-module, then the module M is automorphism-liftable if and only if every P-primary component $M(P)$ of the module M is either the direct sum of isomorphic cyclic uniserial modules of finite length or a uniserial injective module $M(P)$ such that all proper submodules are cyclic and form a countable chain*

$$0 = x_0 R \subsetneqq x_1 R \subsetneqq \cdots$$

such that all subsequent factors of this chain are isomorphic simple modules and $M(P)/x_k R \cong M(P)/x_{k+1} R$ for all $k = 0, 1, 2, \ldots$.

Proof The proof follows from Theorem 8.7 and Remark 8.9.   □

8.4 Idempotent-Lifted Modules and π-Projective Modules

A module M is said to be idempotent-lifted if for any epimorphism $h: M \to \overline{M}$ and every idempotent endomorphism $\overline{f}$ of the module $\overline{M}$, there exists an endomorphism f of the module M with $\overline{f}h = hf$. Without loss of generality, we can assume that $\overline{M}$ is an arbitrary factor module of the module M and $h: M \to \overline{M}$ is the natural epimorphism.

We recall that a module M is said to be π-projective if for any two its submodules X and Y with $X + Y = M$, there exist endomorphisms f and g of the module M with $f + g = 1_M$, $f(M) \subseteq X$ and $g(M) \subseteq Y$ (see [134, p. 359]). The idempotent-lifted modules coincide with the π-projective modules; see Lemma 8.12.

It is clear that every endomorphism-liftable module is idempotent-lifted. If R is a uniserial principal right (left) ideal domain with division ring of fractions Q that is not complete with respect to the $J(R)$-adic topology, then Q is an idempotent-lifted R-module that is not endomorphism-liftable.

The class of idempotent-lifted modules contains projective modules, uniserial modules, local modules and all modules whose factor modules are indecomposable. In particular, all cyclic modules over local rings and all free modules are idempotent-liftable.

Lemma 8.12 *For a module M, the following conditions are equivalent.*

1. *M is an idempotent-liftable module.*
2. *M is a π-projective module.*

Proof $(1) \Rightarrow (2)$. Let $M = X + Y$ be an idempotent-liftable module, $N = X \cap Y, \overline{M} = M/N, \overline{X} = X/N$; let $\overline{f}, \overline{g}_1$ be natural projections of the module $\overline{M} = \overline{X} \oplus \overline{Y}$ onto the components $\overline{X}, \overline{Y}$, respectively; and let and h be the natural epimorphism from the module M onto $\overline{M}$. Since M is an idempotent-liftable module, there are two endomorphisms f, g_1 of the module M such that $\overline{f}h = hf, \overline{g}_1h = hg_1$. It is easy to see that $f(M) \subseteq X, g_1(M) \subseteq Y$. Since $\overline{f} + \overline{g}_1$ coincides with the identity mapping on $\overline{M}$, we have $(1_M - f - g_1)M \subseteq N$. We set $g = 1_M - f$. Then $f(M) \subseteq X, g(M) \subseteq (1 - f - g_1)M + g_1(M) \subseteq N + Y = Y$.

$(2) \Rightarrow (1)$. Let $\overline{M} = M/N$ be an arbitrary factor module of the module M, $h: M \to \overline{M}$, the natural epimorphism, $\overline{f}$ an idempotent endomorphism of the module $\overline{M}, \overline{X} = X/N = \overline{f}(\overline{M}), \overline{Y} = Y/N = (1 - \overline{f})(\overline{M})$, where X, Y are complete pre-images of the modules $\overline{X}$ and $\overline{Y}$ in M, respectively. Then $\overline{M} = \overline{X} \oplus \overline{Y}, M = X + Y$ and $N = X \cap Y$. Since M is a π-projective module, there exist homomorphisms $f: M \to X$ and $g: M \to Y$ with $f + g = 1_M$. Then $\overline{f}h = hf$, and M is an idempotent-liftable module. □

A module M is said to be **automorphism-extendable** if every automorphism of any its submodule can be extended to an endomorphism of the module M. A module M is said to be **idempotent-extendable** if every idempotent endomorphism of any its submodule can be extended to an endomorphism of the module M.

Lemma 8.13 *If R is a ring with 2 invertible in R and M is a right R-module, then every idempotent endomorphism of the module M is the sum of two automorphisms of the module M; in particular, every automorphism-liftable (resp., automorphism-extendable) right R-module is idempotent-liftable (resp., idempotent-extendable).*

Proof Let f be an idempotent endomorphism of the module M. Then $M = X \oplus Y$, where $X = f(M)$ and $Y = (1_M - f)(M)$. We denote by u an automorphism of the module M such that $u(x + y) = x - y$ for $x \in X$, $y \in Y$. Then $f = 1/2 \cdot 1_M + 1/2 \cdot u$, where $1/2 \cdot 1_M$ and $1/2 \cdot u$ are automorphisms of the module M. □

Theorem 8.14 *[114, theorem 1] A module M over a non-primitive Dedekind prime ring R is π-projective if and only if one of the following three conditions holds:*

1. *M is a torsion module such that every primary component is either an indecomposable injective module or the direct sum of isomorphic cyclic modules of finite length.*
2. *$M = T \oplus F$, where T is a nonzero injective torsion module such that every primary component is an indecomposable module and F is a nonzero finitely generated projective module.*
3. *M is a projective module or there exist two positive integers k and n such that the ring R is isomorphic to the ring of all $k \times k$ matrices D_k over some uniserial principal right (left) ideal domain D, $M = X \oplus Y$; X is the finite direct sum of nonzero injective indecomposable torsion-free modules $X_1, \ldots, X_n$; Y is a finitely generated projective module; and either $n = 1$ or $n \geq 2$ and D is a complete domain.*

Theorem 8.15 *[105] Let R be a non-primitive hereditary Noetherian prime ring and M be a right R-module.*

1. *If M is a torsion module, then M is endomorphism-liftable if and only if every primary component of the module M is either an indecomposable injective module or a projective $R/r(M)$-module.*
2. *If M is a mixed module, then M is endomorphism-liftable if and only if $M = T \oplus F$, where T is a torsion injective module such that all primary components are indecomposable and F is a finitely generated projective module.*
3. *If M is a torsion-free module, then M is endomorphism-liftable if and only if either M is projective or R is a special ring with classical ring of fractions Q and $M = E^n \oplus F$, where E is a minimal right ideal of the ring Q, F is a finitely generated projective module, and n is a positive integer.*

Theorem 8.16 *[104, theorem 2] If R is a non-primitive hereditary Noetherian prime ring, then a right R-module M is quasi-projective if and only if either M is a torsion module and every its primary component is a projective $R/r(M)$-module, or M is projective, or R is a special ring and $M = E \oplus F$,* *where E is an injective finite-dimensional torsion-free module and F is a finitely generated projective module.*

Lemma 8.17 *If M is an automorphism-liftable module with local endomorphism ring* End(M), *then M is an endomorphism-liftable module.*

Proof Let $h\colon M \to \overline{M}$ be an epimorphism and $\overline{f}$, an endomorphism of the module $\overline{M}$. If $\overline{f}$ is an automorphism of the module $\overline{M}$, then it follows from the assumption that there exists an endomorphism f of the module M that $\overline{f}h = hf$.

We assume that $\overline{f}$ is not an automorphism of the module M. By assumption, the ring End(M) is local. Therefore, $1_{\overline{M}} - \overline{f}$ is an automorphism of the module $\overline{M}$. Since M is an automorphism-liftable module, there exists an endomorphism g of the module M such that $(1_{\overline{M}} - \overline{f})h = hg$. We denote by f the endomorphism $1 - g$ of the module M. Since $h - 1_{\overline{M}}h = 0$, we have

$$hf = h - hg = h - 1_{\overline{M}}h + \overline{f}h = \overline{f}h.$$

Therefore, M is an endomorphism-liftable module.   □

Theorem 8.18 *A module M over a non-primitive Dedekind prime ring R is an automorphism-liftable, idempotent-liftable module if and only if one of the following three conditions holds:*

1. *M is a torsion module and every primary component of the module M is either an indecomposable injective module or a projective $R/r(M)$-module.*
2. *M is a mixed module and $M = T \oplus F$, where T is a torsion injective module such that all primary components are indecomposable and F is a finitely generated projective module.*
3. *M is a torsion-free module and either M is projective or R is a special ring with classical ring of fractions Q and $M = E \oplus F$, where E is a minimal right ideal of the ring Q, F is a finitely generated projective module and n is a positive integer.*

Proof If one of the conditions (1), (2) or (3) holds, then it follows from Theorem 8.15 that M is an endomorphism-liftable module. In particular, M is an automorphism-liftable, idempotent-liftable module.

Now let M be an automorphism-liftable, idempotent-liftable module. By Lemma 8.12, M is a π-projective module. By Theorem 8.14, either one of the conditions (1) and (2) of our theorem holds or the following condition holds: there exist two positive integers k and n such that the ring R is isomorphic to the ring of all $k \times k$ matrices D_k over some uniserial Noetherian domain D, $M = X \oplus Y$; X is a finite direct sum of nonzero injective indecomposable torsion-free modules $X_1, \ldots, X_n$; Y is a finitely generated projective module and either $n = 1$ or $n \geq 2$ and D is a complete domain.

If $n \geq 2$ and D is a complete domain, then R is a special ring, which is required.

Without loss of generality, we need only consider the case where $n = 1$ and X_1 is a non-projective automorphism-liftable, idempotent-liftable, injective, torsion-free, non-zero module. By Theorem 8.15 and lemma 8.17, the ring R is special. $\square$

Theorem 8.19 *[90] If R is a non-primitive Dedekind prime ring with simple Artinian ring of fractions Q, then the following conditions are equivalent.*

1. *Q_R is a quasi-projective module.*
2. *$R = D_n$, where D is a local principal right (left) domain that is complete in the topology defined by powers of the Jacobson radical $J(D)$.*

In this case, R is a Dedekind prime ring, $J(R)$ is a maximal ideal, and $_R Q$ is a quasi-projective module.

A ring R is called a special Dedekind prime ring if R satisfies the conditions of the preceding theorem.

The third main result of this chapter is the next theorem, where all automorphism-liftable R-modules are described when R is a non-primitive Dedekind prime ring with 2 invertible in R.

Theorem 8.20 *Let R be a non-primitive Dedekind prime ring with $1/2 \in R$. A right R-module M is automorphism-liftable if and only if one of the following three conditions holds:*

1. *If M is a torsion module, then M is endomorphism-liftable if and only if every primary component of the module M is either an indecomposable injective module or a projective $R/r(M)$-module.*
2. *If M is a mixed module, then M is endomorphism-liftable if and only if $M = T \oplus F$, where T is a torsion injective module such that all primary components are indecomposable and F is a finitely generated projective module.*

3. *If M is a torsion-free module, then M is endomorphism-liftable if and only if either M is projective or R is a special ring with classical ring of fractions Q and M = E ⊕ F, where E is a minimal right ideal of the ring Q, F is a finitely generated projective module and n is a positive integer.*

Proof By Lemma 8.13, every automorphism-liftable right R-module is an idempotent-liftable module. Therefore, Theorem 8.20 follows from Theorem 8.18. □

Notes

The results of this chapter are taken from [4] and [127].

9

Open Problems

We conclude the book by presenting a list of open problems. These are of varying levels of difficulty, and we hope that these problems will stimulate interest in the topic among young and senior researchers alike.

A right R-module M is called algebraically compact if every finitely soluble family of linear equations over R in M has a simultaneous solution. It was proved by Warfield that a module is pure-injective if and only if it is algebraically compact [132]. This naturally leads us to raise the following two problems.

Problem 1 *Describe modules that are invariant under endomorphisms or automorphisms of their pure-injective envelopes in terms of systems of linear equations.*

Problem 2 *Study modules that are invariant under endomorphisms or automorphisms of their pure-injective envelopes in terms of their finitely definable subgroups.*

A module M is invariant under endomorphisms (resp., automorphisms) of its injective envelope if and only if any homomorphism (resp., monomorphism) from a submodule to it extends to an endomorphism of M. There is a lack of similar characterization for modules that are invariant under endomorphisms or automorphisms of their pure-injective or cotorsion envelopes. We raise the following problem to address this issue.

Problem 3 *Find characterizations for modules that are invariant under endomorphisms or automorphisms of their pure-injective or cotorsion envelopes in terms of extension of some kind of homomorphisms of their certain submodules.*

Problem 4 *Given a module M, there exists a minimal submodule Q (resp., A) of its injective envelope that is endomorphism invariant (resp., automorphism-invariant). Characterize which modules have a similar property with respect to their pure-injective or cotorsion envelope.*

Ziegler introduced in [140] a notion of partial morphisms related to pure-injectivity and a notion of small extensions of modules based on them. He then proved that for any left R-module M and any monomorphism $u : M \to E$, with E pure-injective, there exists a direct summand E' of E containing M such that the induced extension $M \to E'$ is small. We propose the following problem in connection to this.

Problem 5 *Extend the notion of endomorphism-invariant and automorphism-invariant modules to these extensions and study their endomorphism rings.*

The preceding notions of Ziegler have been recently extended to arbitrary exact categories in [16]. This motivates us to raise the following problem.

Problem 6 *Give a notion of endomorphism-invariant and automorphism-invariant objects in this more general framework of additive exact categories and obtain their main properties.*

In 1974, Boyle conjectured that if R is a ring such that every quasi-injective right R-module is injective, then R must be a right hereditary ring [12]. This conjecture is still unresolved. The conjecture proposed in what follows is inspired by Boyle's conjecture, and it may be viewed as a weaker version of Boyle's conjecture.

Conjecture 7 Let R be a ring such that every automorphism-invariant right R-module is injective. Then R must be a right hereditary ring.

Clearly, a counterexample to the preceding conjecture would be a counterexample to Boyle's conjecture as well.

Problem 8 *Suppose R is a ring such that it is automorphism-invariant both as a right R-module and as a left R-module. Must R be directly finite?*

The motivation behind the question is the fact that if R is both right and left self-injective, then R is known to be directly finite. In fact, Utumi proved a much stronger result that if a ring R is both right and left CS, then R is directly finite.

Problem 9 *Is every torsion automorphism-invariant module over an HNP ring a quasi-injective module?*

Jain and Singh in [64, p. 362], gave an example of a module over an HNP ring, claiming that this module is pseudo-injective (equivalently, automorphism-invariant) but not quasi-injective. It turns out that this example is incorrect. Let us choose as the field Φ in that example the field $\mathbb{C}$ of complex numbers. Then B is a $\mathbb{C}$-algebra, and thus, $\text{End}_B(M)$ is also a $\mathbb{C}$-algebra. Therefore, $\text{End}(M)$ cannot have a homomorphic image isomorphic to $\mathbb{F}_2$. This means that M must be quasi-injective by Theorem 2.32, thus contradicting the assertion in [64, p. 362] that M is not quasi-injective. Since this alleged counterexample is not correct, the preceding question becomes relevant.

If the answer to the preceding question turns out to be true, then it would be worth looking at the following question:

Problem 10 *Is every automorphism-invariant module over an HNP ring, quasi-injective?*

In Chapter 6, we have seen that the Schröder–Bernstein problem has a positive solution for automorphism-invariant modules and pure-automorphism-invariant modules. So we ask the following.

Problem 11 *Do flat modules that are invariant under automorphisms of their cotorsion envelopes satisfy the corresponding Schröder–Bernstein property?*

Faith asked whether a ring in which the injective envelope of every finitely generated right module is Σ-injective is right Noetherian. We ask the next two questions as generalizations of it.

Problem 12 *Assume that the quasi-injective envelope of any finitely generated right R-module is Σ-quasi-injective. Is R right Noetherian?*

Problem 13 *Assume that any automorphism-invariant right R-module that contains an essential finitely generated submodule is Σ-quas-injective (equivalently, Σ-automorphism-invariant). Is R right Noetherian?*

Problem 14 *(a) Let $u : M \to X$ be an envelope, and assume that M is $\mathcal{X}$-automorphism-invariant (or $\mathcal{X}$-endomorphism-invariant). Assume that $\text{End}(M)$ is semiregular. When is $\text{End}(X)$ also semiregular?*

(b) Assume that $\text{End}(M)$ is semiregular and right self-injective modulo its Jacobson radical. When does $\text{End}(X)$ satisfy the same properties?

We have seen that an automorphism-invariant module is a C2 module, it satisfies the exchange property and it is also a clean module. However, not every C2 module with the exchange property is a clean module. For example, if R is a von Neumann regular ring, then R as a right R-module is a C2 module, and it satisfies the exchange property; however, R need not be clean (see, for instance, Bergman's example mentioned in Chapter 2). This prompts the following question.

Problem 15 *Let M be a C2 module such that M satisfies the exchange property. What additional conditions ensure that M is a clean module?*

We have outlined in Section 1.11 the main properties of coprime pairs. As stated in Theorem 1.129, a ring is an exchange ring if and only if, for any left coprime pair (a, b), there exists a minimal left coprime pair $(e, 1 - e)$ that is smaller than or equal to it. On the other hand, we know by Theorem 1.130 that if R is a right cotorsion ring, then there exists a minimal left coprime pair $(e, 1 - e)$ below any descending chain of left coprime pairs. By means of it, it was proved in [44] that any right cotorsion ring is not only an exchange ring but a semiregular ring satisfying that $R/J(R)$ is right injective.

Problem 16 *Adapt the arguments of the proof of Theorem 1.130 to construct idempotents over new classes of rings.*

Problem 17 *Characterize the structure of exchange rings in terms of their descending chains of coprime pairs.*

In Theorem 4.53, we have seen that if R is a ring over which every cyclic right R-module is automorphism-invariant, then $R \cong S \times T$, where S is a semisimple Artinian ring, and T is a right square-free ring such that, for any two closed right ideals X and Y of T with $X \cap Y = 0$, $\mathrm{Hom}(X, Y) = 0$. As a natural generalization, it would be worth looking at the next problem.

Problem 18 *Characterize rings over which all proper cyclic modules are automorphism-invariant.*

The problem dual to this is given next.

Problem 19 *Characterize rings over which all cyclic right modules are dual automorphism-invariant.*

In [62], Ivanov studied fQ-rings; i.e. rings whose finitely generated right ideals are quasi-injective. In particular, some structural theorems on self-injective regular rings were generalized. In connection to these results, the following problem naturally arises.

Problem 20 *Characterize rings in which every finitely generated right ideal is automorphism-invariant.*

In [38] and [65], the authors study the problem of describing the structure of semiperfect rings whose right ideals are quasi-projective. The following question is a natural extension of this problem.

Problem 21 *Characterize rings in which every right ideal is dual automorphism-invariant.*

If all right ideals of a ring R are quasi-projective, then R satisfies the preceding property. In particular, all right hereditary rings or rings with quasi-injective cyclic right modules are such rings.

A ring R is called a right Σ-a ring (resp., Σ-q ring) if every right ideal of R is a finite direct sum of automorphism-invariant (resp., quasi-injective) right ideals. Jain, Singh and Srivastava introduced Σ-q rings in [66], and Σ-a rings were introduced by Singh and Srivastava in [94]. In [94], the following problem was posed.

Problem 22 *Characterize rings in which every right ideal is a finite direct sum of automorphism-invariant right ideals.*

In addition to the rings from Problem 21, every Artinian local ring with square-zero radical satisfies the preceding property.

Problem 23 *Let R be right automorphism-invariant ring and G be a finite group. Is the group ring $R[G]$ right automorphism-invariant?*

The preceding problem is inspired by the well-known result due to Connell that if R is a right self-injective ring and G is a finite group, then the group ring $R[G]$ is right self-injective [15].

In [47], it is shown that if K is any field with more than two elements, then the group algebra $K[G]$ is automorphism-invariant if and only if G is a finite group. We would like to ask the following.

Problem 24 *If $\mathbb{F}_2[G]$ is automorphism-invariant, must G be a finite group?*

We believe that the answer to both the preceding questions is "yes." Finally, we would like to propose the following conjecture.

Conjecture 25 The group ring $R[G]$ is right automorphism-invariant if and only if R is right automorphism-invariant and G is finite.

Problem 26 *Let R be a ring such that each cyclic right R-module is endomorphism-extendable (resp., automorphism-extendable). Is it true that each cyclic left R-module is endomorphism-extendable (resp., automorphism-extendable)?*

Problem 27 *Characterize rings R over which each cyclic right module is automorphism-extendable.*

Problem 28 *Characterize rings R over which each endomorphism-extendable module is strongly endomorphism-extendable.*

Clearly, every strongly automorphism-extendable module is automorphism-extendable. So far, we are not aware of an example of automorphism-extendable module that is not strongly automorphism-extendable. So we propose the next problem.

Problem 29 *Is there an automorphism-extendable module that is not strongly automorphism-extendable?*

We have discussed many situations in which an automorphism-extendable right R-module M is strongly automorphism-extendable. For instance, if R is right Noetherian or M is semi-Artinian, then an automorphism-extendable right R-module M is strongly automorphism-extendable. In the view of this, we would like to propose the next problem.

Problem 30 *Characterize rings R over which each automorphism-extendable R-module is strongly automorphism-extendable.*

Problem 31 *Let M be an endomorphism-extendable R-module, which is an essential extension of a quasi-injective R-module. Must M be quasi-injective?*

Problem 32 *Let M be an automorphism-extendable R-module, which is an essential extension of an automorphism-invariant module. Is M necessarily strongly automorphism-invariant?*

Problem 33 *Let M be an endomorphism-extendable R-module, which is an essential extension of a semisimple R-module. Must M be quasi-injective?*

Problem 34 *Let M be an automorphism-extendable R-module, which is an essential extension of a semisimple module. Is M necessarily strongly automorphism-invariant?*

Problem 35 *Let R be a ring such that each cyclic right R-module is endomorphism-extendable. Is it true that the module R_R is strongly endomorphism-extendable? Is it true that each cyclic right R-module is strongly endomorphism-extendable?*

Problem 36 *Let R be a ring such that each cyclic right R-module is automorphism-extendable. Is it true that the module R_R is strongly* *automorphism-extendable? Is it true that each cyclic right R-module is strongly automorphism-extendable?*

By Theorem 7.18, a semi-Artinian module M is (strongly) automorphism-extendable if and only if M is automorphism-invariant. In [3], it was proved that a right module M over a right perfect ring is automorphism-coinvariant if and only if M is strongly automorphism-liftable. The following question naturally arises from these results.

Problem 37 *Let R be a right perfect ring. Is it true that every automorphism-liftable right R-module is dual automorphism-invariant?*

Problem 38 *Are there automorphism-liftable R-modules that are not endomorphism-liftable?*

If R is a non-primitive Dedekind prime ring and $\frac{1}{2} \in R$, then the automorphism-liftable R-modules coincide with the endomorphism-liftable R-modules by Theorems 8.20 and 8.15. We believe there should be automorphism-liftable modules that are not endomorphism-liftable, but we do not know any such example yet.

By Lemma 8.13, every automorphism-liftable R-module is idempotent-liftable provided R contains $\frac{1}{2}$. This raises the following question.

Problem 39 *Are there automorphism-liftable modules that are not idempotent-liftable?*

Let F be a field, $R = F[[x]]$ be the formal power series ring, and let $M = F((x))$ be the Laurent series ring. Then R is a commutative non-primitive Dedekind domain and M is a strongly automorphism-liftable R-module that is not singular.

Problem 40 *For a non-primitive Dedekind prime ring R, describe automorphism-liftable R-modules and strongly automorphism-liftable R-modules.*

The answer to this question is related to the study of invertible elements of the ring R and automorphisms of cyclic R-modules.

References

[1] A. N. Abyzov, M. T. Kosan and T. C. Quynh, On (weakly) co-Hopfian automorphism-invariant modules, *Comm. Algebra*, (2020), doi: 10.1080/00927872.2020.1723613.

[2] A. N. Abyzov, V. T. Le, C. Q. Truong and A. A. Tuganbaev, Modules coinvariant under the idempotent endomorphisms of their covers, *Sib. Math. J.*, 60, 6 (2019), 927–939.

[3] A. N. Abyzov, T. C. Quynh and D. D. Tai, Dual automorphism-invariant modules over perfect rings, *Sib. Math. J.*, 58, 5 (2017), 743–751.

[4] A. N. Abyzov and C. Q. Truong, Lifting of automorphisms of factor modules, *Comm. Algebra*, 46, 11 (2018), 5073–5082.

[5] A. Alahmadi, N. Er and S. K. Jain, Modules which are invariant under monomorphisms of their injective hulls, *J. Aust. Math. Soc.*, 79 (2005), 349–360.

[6] A. Alahmadi, A. Facchini and N. K. Tung, Automorphism-invariant modules, *Rend. Sem. Mat. Univ. Padova*, 133 (2015), 241–259.

[7] F. W. Anderson and K. R. Fuller, *Rings and Categories of Modules*, Graduate Texts in Mathematics, 13, Springer-Verlag, New York, 1992.

[8] P. Ara, K. R. Goodearl, K. C. ÓMera and E. Pardo, Diagonalization of matrices over regular rings, *Linear Algebra Appl.*, 265 (1997), 147–163.

[9] H. Bass, Finitistic dimension and a homological generalization of semi-primary rings, *Trans. Amer. Math. Soc.*, 95, 3 (1960), 466–488.

[10] K. I. Beidar, S. K. Jain, The structure of right continuous right π-rings, *Comm. Algebra*, 32, 1 (2004), 315–332.

[11] L. Bican, R. El Bashir, E. Enochs, All modules have flat covers, *Bull. London Math. Soc.*, 33 (2001), 385–390.

[12] A. K. Boyle, Hereditary QI-rings, *Trans. Amer. Math. Soc.*, 192 (1974), 115–120.

[13] S. Breaz, G. Calugareanu and P. Schultz, Subgroups which admit extensions of homomorphisms, *Forum Math.*, 27, 5 (2015), 2533–2549.

[14] R. T. Bumby, Modules which are isomorphic to submodules of each other, *Arch. der Math.* 16 (1965), 184–185.

[15] I. Connell, On the group ring, *Canad. J. Math.*, 15 (1963), 650–685.

[16] M. Cortés-Izurdiaga, P. A. Guil Asensio, B. Kaleboğaz and A. K. Srivastava, Ziegler partial morphisms in additive exact categories, *Bull. Math. Sci.*, (2020), doi: 10.1142/S1664360720500125.

[17] W. Crawley-Boevey, Locally finitely presented additive categories, *Comm. Algebra* 22 (1994), 1641–1674.

[18] V. P. Camillo, Distributive modules, *J. Algebra*, 36, 1 (1975), 16–25.

[19] V. P. Camillo and H.-P. Yu, Exchange rings, units and idempotents, *Comm. Algebra*, 22, 12 (1994), 4737–4749.

[20] V. P. Camillo, D. Khurana, T. Y. Lam, W. K. Nicholson and Y. Zhou, Continuous modules are clean, *J. Algebra*, 304, 1 (2006), 94–111.

[21] P. Crawley and B. Jónsson, Refinements for infinite direct decompositions of algebraic systems, *Pacific J. Math.*, 14 (1964), 797–855.

[22] S. E. Dickson and K. R. Fuller, Algebras for which every indecomposable right module is invariant in its injective envelope, *Pacific J. Math.*, 31, 3 (1969), 655–658.

[23] J. Dieudonné, La théorie de Galois des anneux simples et semi-simples, *Comment. Math. Helv.*, 21 (1948), 154–184.

[24] N. V. Dung, D. V. Huynh, P. F. Smith and R. Wisbauer, *Extending Modules*, John Wiley and Sons Inc., New York, 1994.

[25] B. Eckmann and A. Schopf, Uber injektive moduln, *Archiv der Math.*, 4, 2, (1953), 75–78.

[26] D. Eisenbud and P. Griffith, Serial rings, *J. Algebra*, 17 (1971), 389–400.

[27] E. Enochs, Injective and flat covers, envelopes and resolvents, *Israel J. Math.* 39, (1981), 189–209.

[28] E. Enochs, J. R. García Rozas, O. M. G. Jenda and L. Oyonarte, Compact coGalois groups, *Math. Proc. Cambridge Phil. Soc.*, 128, 2 (2000), 233–244.

[29] E. Enochs, S. Estrada and J. R. García Rozas, Galois and coGalois groups associated with cotorsion theories, *Houston J. Math.*, 32, 3 (2006), 651–663.

[30] E. Enochs, S. Estrada, J. R. García Rozas and L. Oyonarte, Flat covers in the category of quasi-coherent sheaves over the projective line, *Comm. Algebra*, 32, 4 (2004), 1497–1508.

[31] N. Er, Rings whose modules are direct sums of extending modules, *Proc. Amer. Math. Soc.*, 137, 2 (2009), 2265–2271.

[32] N. Er, S. Singh and A. K. Srivastava, Rings and modules which are stable under automorphisms of their injective hulls, *J. Algebra*, 379 (2013), 223–229.

[33] C. Faith, *Algebra: Rings, Modules and Categories I*, Springer, Berlin, New York, 1973.

[34] C. Faith, *Algebra II*, Springer, Berlin, New York, 1976.

[35] C. Faith and Y. Utumi, Quasi-injective modules and their endomorphism rings, *Arch. Math.*, 15 (1964), 166–174.

[36] L. Fuchs, On quasi-injective modules, *Ann. Sculoa Norm. Sup. Pisa*, 23 (1969), 541–546.

[37] R. Göbel and J. Trlifaj, *Approximations and Endomorphism Algebras of Modules, Volume 1, Approximations*. 2nd revised and extended ed. de Gruyter Expositions in Mathematics, 41. Walter de Gruyter GmbH & Co. KG, Berlin, 2012.

[38] S. C. Goel and S. K. Jain, Semiperfect rings with quasi-projective left ideals, *Math. J. Okayama*, 19 (1976), 39–43.

[39] K. R. Goodearl, *Ring Theory*, Marcel Dekker, New York, 1976.

[40] K. R. Goodearl, *Von Neumann Regular Rings*, Krieger Publishing Company, Malabar, Florida, 1991.

[41] K. R. Goodearl and R. B. Warfield, *An Introduction to Noncommutative Noetherian rings*, London Mathematical Society Student Texts 16, Cambridge University Press, Cambridge (1989).

[42] R. Gordon and J. C. Robson, Krull dimension, memoirs. *Amer. Math. Soc.*, 133 (1973), 1–78.

[43] W. T. Gowers, A solution to the Schroeder–Bernstein problem for Banach spaces, *Bull. London Math. Soc.*, 28 (1996), 297–304.

[44] P. A. Guil Asensio and I. Herzog, Left cotorsion rings, *Bull. London Math. Soc.*, 36 (2004), 303–309.

[45] P. A. Guil Asensio and I. Herzog, Sigma-cotorsion rings, *Adv. Math.*, 191, 1 (2005), 11–28.

[46] P. A. Guil Asensio and A. K. Srivastava, Automorphism-invariant modules satisfy the exchange property, *J. Algebra*, 388 (2013), 101–106.

[47] P. A. Guil Asensio and A. K. Srivastava, Additive unit representations in endomorphism rings and an extension of a result of Dickson and Fuller, *Ring Theory and Its Applications, Contemp. Math., Amer. Math. Soc.*, 609 (2014), 117–121.

[48] P. A. Guil Asensio and A. K. Srivastava, Automorphism-invariant modules, Noncommutative rings and their applications, *Contemp. Math., Amer. Math. Soc.*, 634 (2015), 19–30.

[49] P. A. Guil Asensio, D. Keskin Tütüncü and A. K. Srivastava, Modules invariant under automorphisms of their covers and envelopes, *Israel J. Math.*, 206 (2015), 457–482.

[50] P. A. Guil Asensio, D. Keskin Tütüncü and A. K. Srivastava, Modules invariant under monomorphisms of their envelopes, *Contemporary Math., Amer. Math. Soc.*, volume 715 (2018), 171–179.

[51] P. A. Guil Asensio, D. Keskin Tütüncü, Berke Kaleboğaz and A. K. Srivastava, Modules which are coinvariant under automorphisms of their projective covers, *J. Algebra*, 466 (2016), 147–152.

[52] P. A. Guil Asensio, Berke Kaleboğaz and A. K. Srivastava, Schröder–Bernstein problem for modules, *J. Algebra*, 498 (2018), 153–164.

[53] P. A. Guil Asensio, T. C. Quynh and A. K. Srivastava, Additive unit structure of endomorphism rings and invariance of modules, *Bull. Math. Sci.*, 7 (2017), 229–246.

[54] R. Guralnick and C. Lanski, Pseudosimilarity and cancellation of modules, *Linear Algebra Appl.*, 47 (1982), 111–115.

[55] D. Handelman, Strongly semiprime rings, *Pacific J. Math.*, 60, 1 (1975), 115–122.

[56] D. Handelman, Perspectivity and cancellation in regular rings, *J. Algebra*, 48 (1977), 1–16.

[57] D. Handelman and J. Lawrence, Strongly prime rings, *Trans. Amer. Math. Soc.*, 211 (1975), 209–223.

[58] A. Hattori, A foundation of torsion theory for modules over general rings, *Nagoya Math. J.*, 17 (1960), 147–158.

[59] G. Hochschild, Automorphisms of simple algebras, *Trans. Amer. Math. Soc.*, 69 (1950), 292–301.

[60] B. Huisgen-Zimmermann and W. Zimmermann, Algebraically compact rings and modules, *Math. Z.*, 161 (1978), 81–93.

[61] B. Huisgen-Zimmermann and W. Zimmermann, Classes of modules with the exchange property, *J. Algebra*, 88(2) (1984), 416–434.

[62] G. Ivanov, On a generalisation of self-injective von Neumann regular rings, *Proc. Amer. Math. Soc.*, 124, 4 (1996), 1051–1060.

[63] S. K. Jain and S. Singh, On pseudo-injective modules and self-pseudo-injective rings, *J. Math. Sci.*, 2 (1967), 23–31.

[64] S. K. Jain and S. Singh, Quasi-injective and pseudo-injective modules, *Canadian Math. Bull.*, 18, 3 (1975), 359–366.

[65] S. K. Jain and S. Singh, Rings with quasi-projective left ideals, *Pacific J. Math.*, 60 (1975), 169–181.

[66] S. K. Jain, S. Singh and A. K. Srivastava, On Σ-q rings, *J. Pure Appl. Algebra*, 213(6) (2009), 969–976.

[67] S. K. Jain, A. K. Srivastava and A. A. Tuganbaev, *Cyclic Modules and the Structure of Rings*, Oxford University Press, Oxford, 2012.

[68] S. Janakiraman Skew projective Abelian groups, *Indag. Math.*, 76, 3 (1973), 233–236.

[69] A. V. Jategaonkar, Jacobson's conjecture and modules over fully bounded noetherian rings, *J. Algebra*, 30 (1974), 103–121.

[70] L. Jeremy, Modules et anneaux quasi-continus, *Canad. Math. Bull.*, 17, 2 (1974), 217–228.

[71] R. E. Johnson and F. T. Wong, Quasi-injective modules and irreducible rings, *J. London Math. Soc.*, 36 (1961), 260–268.

[72] I. Kaplansky, Rings of operators, *Mathematics Lecture Note Series*, W. A. Benjamin, New York, 1968.

[73] D. Khurana and A. K. Srivastava, Right self-injective rings in which each element is sum of two units, *J. Alg. Appl.*, 6, 2 (2007), 281–286.

[74] D. Khurana and A. K. Srivastava, Unit sum numbers of right self-injective rings, *Bull. of Aust. Math. Soc.*, 75, 3 (2007), 355–360.

[75] Y. Kuratomi, Decompositions of dual automorphism invariant modules over semiperfect rings, *Sib. Math. J.*, 60, 3 (2019), 490–496.

[76] M. Kutami and K. Oshiro, Strongly semiprime rings and nonsingular quasi-injective modules, *Osaka J. Math.*, 17 (1980), 41–50.

[77] T. Y. Lam, *Exercises in Classical Ring Theory*, Springer, New York, 1995.

[78] J. Lawrence, A singular primitive ring, *Trans. Amer. Math. Soc.*, 45, 1 (1974), 59–62.

[79] T. K. Lee and Y. Zhou, Modules which are invariant under automorphisms of their injective hulls, *J. Algebra Appl.*, 6, 2 (2013), 9 pp.

[80] T. H. Lenagan, Bounded hereditary Noetherian prime rings, *J. London Math. Soc.*, 6 (1973), 241–246.

[81] J. C. McConnell and J. C. Robson, *Noncommutative Noetherian Rings*. New York: Wiley-Interscience, 1987.

[82] A. P. Mishina, On automorphisms and endomorphisms of Abelian groups, *Moscow Univ. Math. Bull.*, 1 (1972), 62–66.

[83] S. H. Mohamed and B. J. Müller, *Continuous and Discrete Modules*, Cambridge University Press, Cambridge, 1990.

[84] W. K. Nicholson, Lifting idempotents and exchange rings, *Trans. Amer. Math. Soc.*, 229, (1977), 269–278.

[85] P. P. Nielsen, Square-free modules with the exchange property, *J. Algebra*, 323, 7 (2010), 1993–2001.

[86] T. C. Quynh, A. N. Abyzov, N. T. T. Ha and T. Yildirim, Modules close to the automorphism-invariant and coinvariant, *J. Algebra Appl.*, 18, 12 (2019), 1950235.

[87] C. Selvaraj and A. S. Santhakumar, Automorphism-liftable modules, *Comment. Math. Univ. Carolin.*, 59, 1 (2018), 35–44.

[88] D. Simson, On pure semi-simple Grothendieck categories I, *Fund. Math.*, 100 (1978), 211–222.

[89] S. Singh, On pseudo-injective modules, *Riv. Mat. Univ. Parma.*, 9 (1969), 59–65.

[90] S. Singh, Quasi-injective and quasi-projective modules over hereditary Noetherian prime rings, *Canad. J. Math.*, 26, 5 (1974), 1173–1185.

[91] S. Singh, Modules over hereditary Noetherian prime rings, *Canad. J. Math.*, 27, 4 (1975), 867–883.

[92] S. Singh and H. Al-Bleehed, Rings with indecomposable modules local, *Beiträge zur Algebra und Geometrie*, 45 (2005), 239–251.

[93] S. Singh and A. K. Srivastava, Dual automorphism-invariant modules, *J. Algebra*, 371 (2012), 262–275.

[94] S. Singh and A. K. Srivastava, Rings of invariant module type and automorphism-invariant modules, *Ring Theory and Its Applications, Contemporary Mathematics, Amer. Math. Soc.*, 609 (2014), 299–311.

[95] L. A. Skornyakov, *Compelemented Modular Lattices and Regular Rings*, Oliver & Boyd, Edinburgh, 1964.

[96] B. Stenström, *Rings of Quotients*, Springer, Berlin, New York, 1975.

[97] W. Stephenson, Modules whose lattice of submodules is distributive, *Proc. London Math. Soc.*, 28, 2 (1974), 291–310.

[98] R. W. Stringall, Endomorphism rings of abelian groups generated by automorphism groups, *Acta. Math.*, 18 (1967), 401–404.

[99] H. Tachikawa, QF-3 rings and categories of projective modules, *J. Algebra*, 28, 3 (1974), 408–413.

[100] M. L. Teply, Pseudo-injective modules which are not quasi-injective, *Proc. Amer. Math. Soc.*, 49, 2 (1975), 305–310.

[101] A. A. Tuganbaev, Structure of modules close to injective, *Sib. Math. J.*, 18, 4 (1977), 631–637.

[102] A. A. Tuganbaev, The structure of modules close to projective modules, *Sbornik: Mathematics*, 35, 2 (1979), 219–228.

[103] A. A. Tuganbaev, Self-injective rings, *Soviet Math. (Iz. VUZ)*, 24, 2 (1980), 87–91.

[104] A. A. Tuganbaev, Quasi-projective modules, *Sib. Math. J.*, 21, 3 (1980), 446–450.

[105] A. A. Tuganbaev, Semiprojective modules, *Sib. Math. J.*, 21, 5 (1980), 725–728.

[106] A. A. Tuganbaev, Poorly injective rings, *Soviet Math. (Iz. VUZ)*, 25, 9 (1981), 59–63.

[107] A. A. Tuganbaev, Rings over which all cyclic modules are skew-injective [In Russian], *Trudi sem. Petrovskogo*, 6 (1981), 257–262.

[108] A. A. Tuganbaev, Integrally closed rings, *Math. USSR-Sb.* 43, 4 (1982), 485–498.

[109] A. A. Tuganbaev, Semiinjective modules, *Math. Notes*, 31, 3 (1982), 230–234.

[110] A. A. Tuganbaev, Small-injective rings, *Russian Math. Surveys*, 37, 5 (1982), 196–197.

[111] A. A. Tuganbaev, Rings with skew-injective cyclic modules [In Russian], *Abelian Groups and Modules*, Tomsk State University, Tomsk (1986), 151–158.

[112] A. A. Tuganbaev, Rings with skew-injective factor rings, *Soviet Math. (Iz. VUZ)*, 35, 1 (1991), 97–108.

[113] A. A. Tuganbaev, *Semidistributive Modules and Rings*, Kluwer Academic Publishers, Dordrecht-Boston-London, 1998.

[114] A. A. Tuganbaev, Modules over bounded Dedekind prime rings, *Sbornik: Mathematics*, 192, 5 (2001), 705–724.

[115] A. A. Tuganbaev, Rings over which all modules are completely integrally closed, *Discrete Math. Appl.*, 21, 4 (2011), 477–497.

[116] A. A. Tuganbaev, Automorphisms of submodules and their extensions, *Discrete Math. Appl.*, 23, 1 (2013), 115–124.

[117] A. A. Tuganbaev, Characteristic submodules of injective modules, *Discrete Math. Appl.*, 23, 2 (2013), 203–209.

[118] A. A. Tuganbaev, Extensions of automorphisms of submodules, *J. Math. Sci. (New York)*, 206, 5 (2015), 583–596.

[119] A. A. Tuganbaev, Automorphism-invariant modules, *J. Math. Sci. (New York)*, 206, 6 (2015), 694–698.

[120] A. A. Tuganbaev, Characteristic submodules of injective modules over strongly prime rings, *Discrete Math. Appl.*, 24, 4 (2014), 253–256.

[121] A. A. Tuganbaev, Automorphism-extendable modules, *Discrete Math. Appl.*, 25, 5 (2015), 305–309.

[122] A. A. Tuganbaev, Modules over strongly prime rings, *J. Algebra Appl.*, 14, 5 (2015), 9 pp.

[123] A. A. Tuganbaev, Automorphism-extendable semi-Artinian modules, *J. Algebra Appl.*, 16, 2 (2017), 1750029, 5 pp.

[124] A. A. Tuganbaev, Automorphism-invariant non-singular rings and modules, *J. Algebra*, 485 (2017), 247–253.

[125] A. A. Tuganbaev, Injective and automorphism-invariant non-singular modules, *Communications in Algebra*, 46, 4 (2018), 1716–172.

[126] A. A. Tuganbaev, Modules over strongly semiprime rings, *Discrete Math. Appl.*, 29, 2 (2019), 143–147.

[127] Askar Tuganbaev, *Automorphism-Liftable Modules, to appear*. Also see (2019), arXiv:1907.00947 [math.RA].

[128] A. A. Tuganbaev, Automorphism-extendable and endomorphism-extendable Modules, *J. Math. Sci. (New York)*, 245, 2 (2020), 234–284.

[129] Y. Utumi, On continuous rings and self injective rings, *Trans. Amer. Math. Soc.*, 118 (1965), 1–11.

[130] P. Vamos, 2-Good rings, *The Quart. J. Math.*, 56 (2005), 417–430.

[131] R. B. Warfield, Jr., Decompositions of injective modules, *Pacific J. Math.*, 31 (1969), 263–276.

[132] R. B. Warfield, Purity and algebraic compactness for modules, *Pacific J. Math.*, 28 (1969), 699–719.

[133] R. B. Warfield, Exchange rings and decompositions of modules, *Math. Ann.*, 199 (1972), 31–36.

[134] R. Wisbauer, *Foundations of Module and Ring Theory*, Gordon and Breach, Philadelphia, 1991.

[135] K. G. Wolfson, An ideal theoretic characterization of the ring of all linear transformations, *Amer. J. Math.*, 75 (1953), 358–386.

[136] L. E. T. Wu and J. P. Jans, On quasi-projectives, *Illinois J. Math.*, 11 (1967), 439–448.

[137] J. Xu, Flat covers of modules, *Lecture Notes in Mathematics*, 1634. Springer-Verlag, Berlin, 1996.

[138] H.-P. Yu, Stable range one for exchange rings, *J. Pure Appl. Algebra*, 98 (1995), 105–109.

[139] D. Zelinsky, Every lnear transformation is sum of nonsingular ones, *Proc. Amer. Math. Soc.*, 5 (1954), 627–630.

[140] M. Ziegler, Model theory of modules, *Ann. Pure Appl. Logic*, 26, 2 (1984), 149–213.

[141] W. Zimmermann, (Σ-) algebraic compactness of rings. *J. Pure Appl. Algebra*, 23 (1982), 319–328.

Index